Routledge Revivals

The Nature of Life

First published in 1961, this book explains the main trends and problems in modern biological thought, at that time. It was based on lectures presented at the University College of the West Indies, Jamaica, in 1960 to members from different faculties and is therefore an accessible guide for all to the subject.

Routledge Revivals

The Nature of Life

[illegible]

The Nature of Life

C. H. Waddington

First published in 1961
by George Allen & Unwin

This edition first published in 2016 by Routledge
2 Park Square, Milton Park, Abingdon, Oxon, OX14 4RN
and by Routledge
711 Third Avenue, New York, NY 10017

Routledge is an imprint of the Taylor & Francis Group, an informa business

Publisher's Note
The publisher has gone to great lengths to ensure the quality of this reprint but points out that some imperfections in the original copies may be apparent.

Disclaimer
The publisher has made every effort to trace copyright holders and welcomes correspondence from those they have been unable to contact.

A Library of Congress record exists under LC control number: 62003348

ISBN 13: 978-1-138-95700-8 (hbk)
ISBN 13: 978-1-315-66539-9 (ebk)
ISBN 13: 978-1-138-95701-5 (pbk)

C. H. WADDINGTON
F.R.S.

Professor of Animal Genetics
University of Edinburgh

The Nature of Life

London
GEORGE ALLEN & UNWIN LTD
RUSKIN HOUSE MUSEUM STREET

FIRST PUBLISHED IN 1961

PRINTED IN GREAT BRITAIN
in 11 *on* 12 *point Bell type*
BY SIMSON SHAND LTD
LONDON, HERTFORD AND HARLOW

PREFACE

THIS book is based on a course of lectures delivered to the staff and students of the University College of the West Indies in Kingston, Jamaica, in the spring of 1960. The College organizes each year a 'Background Series', consisting partly of single lectures given by its own staff and partly of a group given by someone invited from outside, the aim being to offer, to the whole University, an opportunity for members of the different Faculties to learn something of each other's interests and thoughts. I was deeply honoured at being asked to take part in this most valuable part of the university's activities, and also, of course, grateful for the opportunity of visiting that most beautiful part of the world. But I have been only too conscious of the great difficulty of the task the College set me. I had to try to expound the main problems and trends of thought in modern biology to an audience drawn from all Faculties, and of all ages from the youngest freshman to the most senior professors; and at the same time, my colleagues from the biological departments would, I suppose, hope to hear something fresh enough to prevent them being bored. I can hardly expect to have satisfied in equal measure throughout the whole book this great variety of outlook and sophistication; it will be enough if each reader finds the lectures, like the curate's egg, 'good in parts'.

Of the many people who made my stay in Kingston so enjoyable, I should like particularly to thank Dr P. M. Sherlock, the Acting Principal at the time of my visit; Professor A. P. Thornton, Chairman of the Senate Committee which manages this series of lectures; Mr H. W. Springer, the Registrar, and Professor D. M. Steven, who not only made me at home in his department of zoology, but found time to show me some of the magnificent biological and other beauties of Jamaica's forests, mountains and coral reefs.

C.H.W.

CONTENTS

An asterisk in the text refers to a note at the end of the book.

CHAPTER 1

The Natural Philosophy of Life

THERE are almost as many good reasons for doing science as there are scientists. When I was asked to speak about my subject to an audience of scholars most of whom were interested in other fields I had first of all to think what makes biology worthy of their attention. Many of the reasons why biologists actually pursue a subject are too egocentric to make a general appeal. It is fun, for instance, to fiddle about with apparatus in a laboratory; to grow cells in tissue culture; to breed mice or fruit flies in cages or bottles, or to observe the inhabitants of coral reefs or woods. But it is not for everyone that the sheer carrying-on of science is such an attractive pastime. When one thinks of the more serious reasons why all intelligent people should be interested in it they fall into two main kinds. One may think science important either because it enables man to control the world he lives in, or alternatively because it enables him to understand it.

These two ways of looking at science are not, of course, by any means completely alternative and independent of one another. We can scarcely claim to have anything like a satisfying scientific understanding of any natural process until we can feel that we are in a position to control it, or at least to see how it might be controlled, even if we are not actually able to carry out the necessary measures. Scientific theories, in fact, are not simply intellectual constructions which give a coherent account of a certain range of phenomena. Indeed they can never provide a system of explanation as logically water-tight as is given by purely rationalistic theories which invoke special forces—such as those Dr Pangloss was so fond of—to deal with each type of process.

A scientific theory cannot remain a mere structure within the world of logic, but must have implications for action and that in two rather different ways. In the first place, it must involve the consequence that if you do so and so, such and such results will follow. That is to say it must give, or at least offer, the possibility of controlling the process; and secondly—and this is a point not so often mentioned by those who discuss the nature of scientific theories—its value is quite largely dependent on its power of suggesting the next step in scientific advance. In fact at the growing edges of science the stimulating and suggestive theory is often preferred to one which may be less open to criticism but which seems to lead nowhere in particular.

It is probably for this reason that so many scientific theories are expressed in terms of model systems which from a strictly logical point of view would seem to be unduly realistic. For instance, the foundation of the science of genetics occurred when Mendel discovered that from crosses between certain types of parents particular categories of offspring were born in definite proportions which could be stated as simple arithmetic ratios. These numerical results could be accounted for in terms of some kind of hypothetical entity, originally known as a Mendelian factor, which follows certain rules as it passes on from parents to offspring in the process of reproduction. Both in the very early days of genetics some authors, such as William Bateson, and even in more recent times other writers such as Woodger, have emphasized that from the standpoint of pure logic these factors are merely constructions built out of the relative numbers of offspring of the various kinds. Other students of the subject, particularly Morgan, were bold enough to explore the plain-man hypothesis that a factor is not a mere logical construction but is actually a material particle. At the time Morgan took this step it was definitely going beyond what the facts warranted; but the material model he adopted immediately suggested many questions which might seem worth looking into, and it was the investigation of these which led to the whole development of the gene theory and modern genetics.

Similarly Dalton's atomic theory in its day, and the later Bohr-Rutherford planetary construction of atoms, both involved the assumption of models of a more detailed material kind than were strictly necessary, but both suggested numerous questions which science could then proceed to investigate. If the atom is built like a solar system with electrons moving in orbit around a nucleus, we can ask how an electron can move from one orbit to another, or whether it can spin on its axis as, for instance, the world does. Science is often quite ready to tolerate some logical inadequacy in a theory—or even a flat logical contradiction like that between the present particle and wave theories of matter—so long as it finds itself in the possession of a hypothesis which offers both the possibility of control and a guide to worthwhile avenues of exploration.

But although for these reasons one cannot make a sharp distinction between Man's effort to control and to understand the world, there is a considerable difference, at least in emphasis, between these two endeavours. Science which is motivated mainly by the desire to control the world will naturally study those aspects of it, the control of which would be most important in practice. The desire to understand, on the other hand, directs one's attention towards problems which have wide implications concerning the general nature of the world and of Man, even if his ordinary daily life does not urgently demand that he should control them. This might seem a less pressing reason for the development of science than the practical needs of human existence. Historically, however, it was the desire to understand rather than the need to control which led to the development of an orderly and organized body of thought and knowledge. The control of the material basis of existence was at first the province of the technician or craftsman, understanding that of the philosopher. The coming together of these two types of activity produced the first scientists. The predominant part which was played by the wish to understand the world was expressed in the original name of their activity, Natural Philosophy.

As science has developed over the last few centuries it has of course provided us with powers of controlling our environment which were quite unthought of, and would have been quite inconceivable, by the early natural philosophers. So much so indeed that we tend nowadays perhaps somewhat to neglect the natural-philosophical aspect of science, and, when the word is mentioned, to think first of the triumphant manipulations of material things which have given us the automobile, the telephone, drugs, antibiotics, electric light and all the rest, including the atom bomb. But these products of science are essentially developments of the craftsman aspect of its nature. They are enormously important in practical affairs, in fact at the present time our whole life is built on them, and most of us could scarcely survive for a few days or weeks if they all suddenly disappeared; but they are based on enormously specialized and detailed understanding of particular types of process, and there is no very good reason why intelligent people in general should go into them very deeply. We should all, of course, have some inkling of what is meant by such scientific terms as hormones, or enzymes or electromagnetic vibrations or chemical valencies, but there is no reason why everyone should feel called upon to obtain even an outline understanding of modern developments in endocrinology, biochemistry, electrical theory or synthetic chemistry. In this set of lectures I shall not attempt to provide anything of that kind, but shall be concerned almost entirely with the natural philosophical aspects of biology, that is to say its contribution towards Man's attempt to understand nature and his place in the system of living things.

Before we start to discuss the problems of biology it will be as well to consider for a minute the nature of science in general. What sort of activity is it, and by what faculties of man is it produced? Even today there are too many people to whom the word suggests the 'bug-eyed scientist in a white coat', or the animate calculating machine only by courtesy flesh and blood, who spends his time processing data according to the rules of an inflexible and almost super-human logic. But the ordinary scientist, or even the extraordinary one, is neither a 'commissar square' nor a 'beat yogi'. He is not in the slightest a creature that stands outside the ordinary community of mankind. In fact, in many ways he is more than most intellectuals a social being.

It is one of the most important characteristics of science that it is not the creation of one man, or even of a succession of individual men. Basically, in essence, and all through and through, it is something which has been produced by co-operative effort. An individual man can, of course, add a brick to the structure, or even lay out the plan of a new room, but his brick must be added to a wall which others have already partially built, and his new room must join on and communicate with the rest of the whole palace of knowledge. Science builds on and incorporates its past to a much greater extent than do any of the other major cultural activities of mankind. In painting, in poetry and the literary arts generally, the balance between originality and tradition is a very uneasy one. If, in the work of an artist, we see clear references to the work of his predecessors, we are likely to dismiss him as derivative. In science, unless we see that a man knows and respects what his forerunners have produced, we are inclined to find him unconvincing and ignorant. This communal, co-operative nature of scientific endeavour is one of its major sources of strength.

It is important to realize how enormous that strength is. The scientific community is, I am afraid it must be confessed, a subversive organization; and successfully—some people would say too successfully—subversive. In those parts of the world where science has been vigorously pursued it has already destroyed the stability of a way of life which had lasted for many centuries, perhaps from the time of the Greeks and Romans until say three or four generations ago. This is, of course, a platitude. Everybody knows it and lots of people say it. But even that does not prevent it still being the most important fact in the world.

The revolution brought about by science has not been due to the political activities of scientists. No country that I can think of has ever had a government the majority of whose members were trained as scientists; most of the countries we usually consider as civilized have been governed by men who were educated in the arts and humanities. The effects of science have been produced in a more subterranean way. Its effectiveness depends partly on the fact that it has found how to combine the efforts of many individuals into a single whole, but there is another point which provides a further reason for the enormous power of the scientific movement.

Let me offer this rather far-reaching argument for your consideration and criticism. I suggest that science can be defined simply as the application, to questions concerning the external world, of *all* the major faculties which man is capable of exercising. Any complete piece of scientific work starts with an activity essentially the same as that of an artist. It starts by asking a relevant question, and that demands either or both of two things. The first step may be a new awareness of some facet of the world that no one else had previously thought worth attending to. Or the scientific advance may start from some new imaginative idea; ideas, for instance, like that of the quantum of energy, of indeterminacy, or lack of parity, or of a unit of biological heredity, or of the notions of releasers, displacement activities and so on, in which people are now analysing animal behaviour. These germs from which scientific work originates depend on a sensitive receptiveness to the oddity of nature essentially similar to that of the artist. When they are first proposed they often have the same quality of unexpectedness, and perhaps wrong-headedness, as say, cubism, abstract art or atonal music.

In science they have to be immediately followed up by the application of another of man's faculties, that for ratiocination. They have to be formulated in precise and probably mathematical language, and incorporated into a body of logical exposition. And then there comes a third activity, that of manipulation and the devising and carrying out of experiments. This is another stage which the scientist shares with the artist, but the latter, of course, leaves out the phase of logical analysis. And then in science there comes a fourth and final phase of the procedure, and this again is one which is usually much less well-developed in the humanities. In his logical analysis and manipulative experimentation, the scientist is behaving somewhat arrogantly towards nature, trying to force her into his categories of thought or to trick her into doing what he wants. But finally, he has to be humble. He has, as T. H. Huxley put it a century ago, to sit down before the fact like a little child. He has to take his intuition, his logical theory and his manipulative skill to the bar of Nature and see whether she answers yes or no; and he has to abide by the result.

It is because science does all those things one after the other, and because they are done not by isolated individuals but co-operatively by the world community of scientists, that it has added, to our understanding of the world we live in, and indeed of ourselves, so much more than many centuries of reliance on pure intuition or pure intellect or on the kind of irrational activity found in magical systems.

Living organisms are, of course, much more complicated than the non-living things which man encounters in his surroundings. Biology—the scientific study of living systems—has therefore developed more slowly than sciences such as physics and chemistry, and, as is inevitable in such a social activity as science, has tended to rely on them for many of its basic ideas. These older physical sciences have, on the whole, provided biology with many firm foundations which have been of the greatest value to it, but throughout most of its history biology has found itself faced with a dilemma as to how far its reliance on physics and chemistry should be pushed. Ever since our understanding of the physical world became tolerably satisfactory there have been some biologists who believed that their ultimate object must be to account for all the phenomena of life in terms of processes which can be discovered amongst non-living things. One can perhaps take Descartes as a representative of this point of view at the beginning of the modern period of science. He made a rigid distinction between the subjective phenomena of thought and feeling—the *res cogitans*—and the objective observable phenomena of the material world—the *res extensa*; and he argued that the latter, which includes the bodies of men and living things in general, must be fully explained in terms of simple material entities and strictly mechanistic principles. Opposed to this 'nothing-but' hypothesis of the *bête machine* there have always been other biologists, of whom one may take Harvey as the representative in Descartes' time, who believed that processes of life involve some vitalistic principle over and above the agencies which operate in the non-living world.

Among the various types of vitalism which have been put forward throughout the centuries it is important to distinguish two main kinds. There is, on the one hand, a thorough-going vitalism, which might be referred to as 'objective vitalism', which claims that the phenomena which we *observe* in living things cannot be fully accounted for by means of the concepts which are adequate for the non-living world. For instance, Driesch was a prominent advocate of this view at the beginning of this century. He argued that the results of certain experiments, in which he cut eggs into fragments and found that each fragment had developed into a complete adult, could not be explained without invoking the activity of some non-material and non-mechanical whole-making active agent, which he called an entelechy. At the other end of the range of the vitalistic theories there is a point of view which could allow that all observable phenomena are potentially explainable in terms of concepts essentially similar to those of physics and chemistry, but which insists that phenomena of self-awareness—which we can never observe objectively, but only experience subjectively in ourselves—are of a radically different nature and cannot be accommodated within the same structure of ideas.

The distinction which such theories draw is perhaps not quite the same as that between Descartes' *res cogitans* and *res extensa*. It is not that the processes of thought as such are completely excluded from the sphere to which a mechanistic explanation might apply. These processes have an aspect which is theoretically open to objective observation, for instance, by some very refined analysis of the electrical currents passing through various nerve cells in the brain. But subjective vitalists would argue that, even if we knew exactly what currents were passing through which cells when a given man thinks of a particular concept, there is no way of jumping the gap which separates our notions of electrical processes in material systems from the subjective awareness of an image, an emotion or an abstract concept. A large proportion of the workings of our nervous system and brain will proceed, as we well know, without our being aware of them; not only such processes of nervous adjustment as those by which we keep our balance when walking, but, as the psycho-analysts have shown, a great deal which has a character much more closely comparable to that which when it is conscious we know as thought. Thus, subjective experience does not seem to be an inevitable correlate of certain types of nervous activity, but appears to be something special; not merely an item which we have not yet allowed for in the conceptual scheme with which we account for observable phenomena, but something which lies in a different realm. Subjective vitalism is difficult to refute; I will have something more to say about it at a later stage in these lectures.

In the last century or so it has been the conflict between mechanism and objective vitalism that has provided the main polarity between which biologists have ranged themselves. Even as recently as my own student days it seemed that the most important question facing a general biologist was to take some position vis-a-vis this great conflict of opinion. But actually at just about that time, or a few years before, the antimony between the two views was resolved, and the whole controversy evaporated. Objective vitalism amounted to the assertion that living things do not behave as though they were nothing but mechanisms constructed of mere material components; but this presupposes that one knows what mere material components are, and what kind of mechanisms they can be built into. In late Victorian times, in the heyday of the physics of Newton and Faraday and the chemistry of Dalton, people had a surprising confidence that they really knew what the world of matter consists of. Atoms and electricity seemed to be well known and completely comprehended things. It appeared quite natural to start from such certainties and to enquire whether they could or could not be used to explain the more complex phenomena observed in the living world. Those who thought they could, or at any rate should, were mechanistic biologists; those who thought they could not, were objective vitalists. A half-way house was represented by the theory of emergent evolution. This argued that when two or more simple entities come together in a particular arrangement they may gain new properties which they previously did not possess. For instance, it was suggested that when sodium and chlorine atoms come together to form common salt there emerges a new property of the compound which was not contained previously in the isolated atoms.

It was, more than anyone else, the philosopher Whitehead who provided the new way of looking at the situation which dehorned the dilemma. The physicists had, by the first two decades of the twentieth century, realized that atoms were not after all the hard billiard balls that their Victorian grandfathers had envisaged, and as the sciences of the inanimate world moved towards a theoretical structure in terms of relativity, quantum theory and indeterminacy, it became less and less plausible to suppose that there is any clear-cut mechanistic system of thought against which vitalism could rise as a reaction. It was Whitehead who put in general terms the point of view that this ancient dilemma arises essentially from seeing the situation upside down. It is not the case that we begin by knowing all about the ultimate constituents of the inorganic world, and can then ask whether they can account for the observable phenomena of biology. Always, whether in physics or in biology, it is from observable phenomena that we have to start; ultimate, or penultimate, or pen-penultimate constituents are what we hope to approach. It was one of the only too numerous aberrations of Victorian self-confidence to think that they knew enough about atoms to provide vitalists and mechanists anything to argue about.

Whitehead's thought was certainly strongly influenced by that of the emergent evolutionists such as Alexander. In fact his ideas about biology can to some extent be regarded as emergent evolution seen from the other end. We start from a variety of observable phenomena, and from these we construct concepts (or models) of simpler entities, by combinations of which we can account for what we have observed. These simpler entities remain, however, mysterious things about which we know no more than we have been able to discover by inspecting the phenomena in which they take part. If we could observe the behaviour of sodium and chlorine only when each is in isolation, and if we regarded these two substances as made up of atoms, we might be able to discover something about these atoms, but not very much. There is no reason why we should expect to be able to become aware of the properties which allow them to combine with one another and form common salt. When this compound is formed, it is not that some new 'emergent' properties appear; it is simply that a new avenue is open to us for discovering a little more about the sodium and chlorine atoms.

Similarly with phenomena of life. When it turns out that certain arrangements of the atoms of carbon, nitrogen, hydrogen, oxygen, etc., exhibit properties which we recognize by the name of enzymes; when other still more complicated arrangements turn out to be able to duplicate themselves identically like the genes in the cell nucleus, or to be able to conduct electrical impulses like nerve cells, or to exhibit the correlated electrical phenomena found in the staggeringly complex systems of nervous cells in the brain; it is completely out of the picture to suggest that we have to add something of a non-mechanistic kind to an already fully comprehended material atom. What we have done is simply to discover something about atoms that we did not know before; namely, that when they are arranged in certain special ways the total complex can exhibit behaviour that we might not have expected at first sight. There is nothing *philosophically* mysterious about this. But still it would be frivolous to consider it unimpressive. Who, seeing a few pieces of glass, metal, plastic and so on, would suspect that they could beat him at chess? Yet we know that, assembled into a computer, they could wipe the floor with any but the world champions. The secret of their performance in this way is architecture, or, to use the Aristotelian term, form. The ultimate constituents of matter—atoms, electrons and so on—hardly become known to us except when they combine into structured entities which have a definite character, and every time they do so they betray to us a little more of their secrets.

Here we come face to face with a paradox which gives rise to a new polarity. It is, I think, the operative polarity in present day biology. As architecture or structural arrangement becomes more complicated, the ultimate constituents are able to conceal more and more of their own character within it. We could make a computer capable of performing a certain type of operation out of a variety of materials, using if you like silver instead of copper, or one chemical variety of plastic rather than another. The architecture itself is more important than the constituents out of which it is built. Is this perhaps also the case, not only for instruments designed by man, but for some of the more complicated entities which occur in nature, such as the living organisms which have been produced by evolution? The answer is not by any means obvious. Clearly a man and a horse, for instance, differ in architecture; and immunology and other methods of analysis demonstrate that they differ also in their basic chemical constitution. Almost certainly these two differences are intimately connected. The architecture must be an expression of the properties of the substances out of which it is built; but does this imply that we can safely neglect the architecture, and concentrate all our endeavours on the exploration of the nature of the chemical differences? Or does the architecture itself offer clues which is worth while to follow up?

There are probably few biologists who would claim that either the architecture or, alternatively, the substance are alone of such overwhelming importance that the other could be wholly neglected. Almost everybody is likely to agree that both requires study. But we confront here a contrast between two ways of tackling scientific problems, which have for many centuries competed with one another over a much wider field than that of biology. On the one side is a mode of approach which is frequently spoken of as atomism. It aims at producing an account of phenomena in terms of a number of self-sufficient entities each of which could exist in complete isolation. The full range of properties of these entities may, of course, only be revealed when they enter into relations with one another. But in an atomistic theory these relations are thought of as external relations, which may reveal, but do not bring into being, the properties, which would persist even if a given entity was isolated from all others. The Daltonian atomic theory of chemistry as it was understood in the middle of the last century is a good example of such a theory. In contrast to these there are 'continuum' theories, that is to say, theories in which the basic constituents of phenomena are not thought of as separate and discrete entities which enter only into external relations with one another. Continuum theories attempt to account for phenomena in terms of sets of relations. Any elementary factors which enter into the expression of such a theoretical structure are no more than nodal points within the set of relations; and their properties depend on, and arise from, the relations into which they enter. The transition from the old billiard ball concept of the atom and electron towards theories phrased in terms of a psi-function, or a probability function spread throughout the whole of space, is a movement from an atomistic theory towards one which is more of the continuum type.

Physics faces the task of studying the ultimate constituents of the world—or at least the most ultimate with which our advancing knowledge has yet brought us in contact. For it, the contrast between atomistic and continuum theories is rather sharply drawn. However, it does not seem possible to decide wholly for one or the other. Physics employs both; an atomistic or particle theory in one context, a wave or continuum theory in others. For biology the contrast is less acute, since there is no 'fundamental biology', in the same sense as there is a fundamental physics; that is to say, no field in which biological analysis comes up against the point at which analysis can proceed no further. We can analyse living organisms into cells, into chromosones, into genes, into sub-genic units such as cistrons and mutational sites; but we can still push the analysis further, although then it turns into chemistry under our hands, and we find that we are discussing perhaps nucleotides or aminoacids. For biology the atom-continuum contrast is not a question about the ultimate way in which the universe is built, but concerns the kind of question which it seems interesting to ask about the particular group of phenomena which constitute the living world. Is each individual gene a self-sufficient atomistic entity, or is it only a focal point in a wider set of relationships in the absence of which it is no longer a gene at all? Shall we try to account for physiology in terms of the activities of isolated cells, or shall we expect to find, or indeed look for, activities of groupings of cells, such as tissues? Or we can put the question in another and very general way. What is most nearly entitled to be called fundamental biology? The end points of biological analysis, such as genes and sub-genic units? Or the most highly developed relationships that living things exhibit, the processes of human civilization?

There is, of course, no one-way answer to any of these alternatives in biology. Like the physicists in their more refined field, biologists have to utilise both atomistic and continuum modes of approach simultaneously. This sets up what might be called a certain tension within the field of biological theory. In my opinion it is this tension which is one of the major springs of the vitality of biological science. In the next few chapters I shall try to expound some of the contexts in present-day biology where the atomistic and the continuum approaches light up various aspects of living processes from different angles.

CHAPTER 2

Innate Potentialities

ONE of the simplest questions to ask about any entity in the natural world is how does it work, what makes the wheels go round? In our intensely mechanical civilization we are surrounded by things which almost force this question on us, since we are rather helpless in the world of practical affairs unless we have at least an inkling of the mode of operation of some of the simpler mechanisms on which we depend so much. In trying to understand living things also the question of how they work is perhaps the first to occur to one. Moreover it is obviously of enormous practical importance. Just as we need, to keep our cars in order, skilled mechanics who understand in detail the workings of different varieties of carburettors and so on, so we require the doctors who look after our bodies to know very precisely the workings of the heart, lungs, intestines and other organs of the body.

The study of the mechanisms by which living things maintain themselves and carry out their day-to-day operations has therefore always comprised the greater part of biology. However, in spite of its volume and extent, it is by no means the whole of biology, or even perhaps the aspect which is of most importance from a general or philosophical point of view. We require to know about living things not only how they work but how they come to be there. Even in the world of machines there are more general and interesting questions than those that we should put to a competent garage mechanic. We can find self-propelling vehicles of all sorts; from double-decker buses to bubble cars; driven by internal combustion engines, by electricity, by steam, by diesels and perhaps by gas turbines; and there are other obviously related mechanisms which move not on the surface of the land but over the sea or in the air. If we want to pass beyond the detail of the working of one kind of machine to a general picture of the whole range of such mechanisms, we can in this case easily see in what general framework the picture can be arranged. All these things—cars, buses, tractors, aeroplanes, ships and so on—have been designed and made by men for particular purposes. They can be ordered in terms of the particular human purposes they were designed to meet.

Admittedly some odd questions will arise. For instance, why is the engine of the ordinary motor car in front, although the drive is applied to the back wheels? The answer is probably at least in part historical, related to the derivation of automobiles from horse-drawn carriages, an evolutionary process whose traces are easy to see in the carriage work of the earliest motor cars. Again we shall find that some features of these mechanisms have acquired their particular character from the requirements of the manufacturing process; certain types of shape of body, for instance, are easier to produce by a mechanical press than others. But if we were to attempt to give an orderly account of man-made self-propelled mechanisms, these questions of historical derivation and of the requirements of manufacturing processes play a comparatively small part in comparison with the influence of the purposes which animated the designer.

When we turn from the mechanisms which man fabricates to those which we find around us as natural living organisms, we find ourselves faced with the same problem of giving a general account of an enormous range of different operating systems, and therefore with the same need to take a broader view than the detailed understanding of how any one system works. We also find ourselves confronted with similar phenomena of historical derivation—in this case the processes of organic evolution; and also with phenomena comparable to those of manufacture—the development of the organism from a fertilized germ to a fully differentiated adult. But there is no human designer of the natural world of living things. We can postulate a super-human designer, a creating God. If he is conceived of as outside the Universe, then his purposes are not open to our understanding as are those of the human designers of vehicles, and it becomes senseless to try to use them as a framework for a general rational understanding of the world as we find it. On the other hand, if the Creator's purposes are expressed in the world, then we have to examine it to discover what they are, since we have no possibility of interrogating him as we can interrogate a human designer. Thus, an appeal to a purposeful Creator as an explanation for the nature of living things either abolishes the possibility of rational biology, or leaves us just where we were before, faced with the need to account for the phenomena of life in terms of the happenings which we can see preceding in front of us.

It is convenient to classify the processes by which living things come into being into the three categories of development, heredity and evolution. For the reasons just stated, processes of these kinds play an enormously more important part in the theory of biology than they do in the theory of man-made mechanisms or of the inorganic world as a whole. It is perhaps the most characteristic feature of biology, and its greatest point of difference from the sciences of physics and chemistry, that it deals with entities which must be envisaged simultaneously on four different time scales. Not only must we study the hour-to-hour or minute-to-minute operations of living things as going concerns, but we cannot leave out of account the slower processes, occurring in period of time comparable to a lifetime, by which the egg develops into the grown-up adult, and finally towards senescence and death. On a longer time scale again, there are phenomena which must be measured in terms of a small number of lifetimes; they are the processes of heredity, by which characteristics of organisms are passed on from parent to offspring. Finally, on the time-scale of many hundreds of generations, there are the slow processes of evolution, by which the character of the individuals in a given population gradually changes, and the population may become split up into two or more different species. Biological theory, which has to cover adequately all four of these time-scales, is of necessity much more complex than that of the physical sciences.

It is a rather atomistic insight which provides perhaps the best Ariadne's thread for those who would penetrate the intricacies of this biological maze. The first principle of an atomistic approach to a problem is to try to catch the entities concerned in their simplest terms. From the time when it is no more than a newly fertilized egg, the mouse is just as much a mouse as the creature you may see disappearing in a hole in the wainscotting; and the human egg is as just as much a man (though perhaps not as much *of* a man) as you are. We have, then, in the fertilized egg something relatively simple and yet comprehensive enough to include, in some way or other, the whole system of a living organism. Understanding of the egg therefore provides the leading clue to the elucidation of the whole complexity of the biological realm.

The fertilized egg, when inspected with the microscope, consists of no more than a small lump of living material which contains two other smaller masses which are in process of fusing together. Living material only occurs in the form of small lumps, which are bounded by some sort of a membrane. Each such lump is known as a cell, and the fertilized egg is also a cell of a particular type. The material which makes up the main body of the cell is known as the cytoplasm. In it there is embodied a smaller mass of a special kind which is known as the nucleus. In the newly fertilized egg the cytoplasm contains at first two such masses, one being the nucleus which was formed in the egg cell as it grew in the body of the mother, and the other being a nucleus which was brought into the egg by the sperm from the father. These two nuclei very soon fuse together to form a single one. The whole body of the newly developing animal will be formed by the gradual multiplication of the egg cell and its contained nucleus; this will grow till it develops into a larger mass made up of innumerable small cells, each containing cytoplasm in which a nucleus is embedded.

It is clear that the character of a newly fertilized egg depends on the nature of the parents from which it was produced. A cock and hen will never give rise to an egg from which a duck will develop. A clue to the understanding of the fertilized egg therefore lies in the study of heredity. The biological discipline concerned with this, usually known nowadays as Genetics, has therefore a key position in the whole edifice of biological theory.*

In spite of its importance, genetics is one of the most recently developed of all the major branches of biology. It was only at the beginning of this century that the nature of the basic principles governing the phenomena of heredity was realized by biologists in general. However, in the sixty years that have elapsed since then, genetics has grown to be one of the most fully developed and logically coherent parts of biological theory. An examination of how it has come into flower, and of the questions which still confront it, will provide a good example of the mutually reinforcing interplay of atomistic and continuum theories which is so characteristic of biology.

The first step in the understanding of heredity is to realize that what a pair of parents donate to their offspring is a set of potentialities, not a set of already formed characteristics. You do not inherit fair or dark hair, blue or brown eyes, from your parents; what you inherit is something which endows you with the capacity for developing eyes of a particular colour under certain particular circumstances, and perhaps a different colour under other circumstances. Even if your parents were both Anglo-Saxons you do not inherit their white skin; you inherit potentialities of such a kind that if you grow up with very weak sunlight your skin will be very fair in colour, while if you are frequently exposed to much stronger sun it will be considerably browner. Nowadays we use the word 'genotype' for the collection of potentialities which are inherited. Contrast this with the 'phenotype', which is the name for the collection of characteristics which an individual actually develops under the particular circumstances in which he happens to grow up. Any one genotype may give rise to many somewhat different phenotypes, corresponding to the different environments in which development occurs.

In the pre-history of genetics—before, that is to say, its development as an orderly science got under way—the genotype was envisaged as having a character roughly comparable to that of a fluid. When two animals mated and produced offspring the hereditary contributions of the two parents were supposed to blend in the newborn creature. In fact the hereditary contribution was often spoken of somewhat metaphorically as 'blood', so that people spoke of animals belonging to certain 'blood lines', and so on. Any such theory of blending inheritance has certain obvious difficulties. For instance, it provides no reason why the sons of any chaste and well-behaved wife should differ from each other, as they in fact obviously do. Then there is a difficulty which particularly impressed Darwin when he was discussing the process of evolution. If the hereditary endowments of the two parents blend in the offspring, then after this has gone on for a few generations all the individuals in an interbreeding

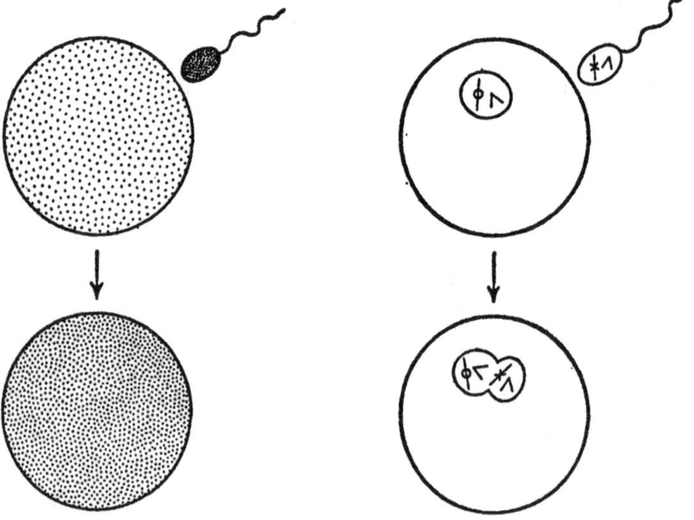

FIG. 1

The two drawings on the left show the old 'blending' theory of heredity; a sperm (dark) unites with an egg (pale) and their characters blend to give an offspring of intermediate type. On the right is the modern theory. The egg contains a nucleus, in which are shown two chromosomes, a V-shaped one and a long rod, on which, at one particular place, is a hereditary factor or gene indicated by a circle. The sperm also contains two similar chromosomes, the rod-chromosome containing a slightly different form of the gene, indicated by a cross. The offspring contains both these different kinds of genes, which come into the same nucleus but do not fuse or blend.

population will have come to resemble one another in some intermediate nondescript average type, just as a mixture of all the paints in a paintbox turns into a muddy brown. As Darwin realized, if this were to happen there would be no individual differences left, and the natural selection of the most effective variants, on which he founded his theory of evolution, would have to come to a stop for lack of anything to work on.

Darwin never saw how to get over this difficulty. The answer was found by Mendel, who actually wrote during Darwin's lifetime but whose work was surprisingly neglected until the beginning of the twentieth century. Mendel rejected entirely the continuum outlook involved in the idea of blending inheritance, and introduced a boldly atomistic theory. He suggested that the hereditary endowment passed on from parent to offspring consists of a number of separate discrete hereditary factors. Each factor is concerned with some particular potentiality for development. The whole complex character of an individual is to be thought of, according to Mendel, resulting from the realization of a number of separate potentialities—to put it in its simplest terms, if you like, one potentiality concerned with eye colour, and another concerned with skin colour, another concerned with the shape of the bones, and so on throughout a whole large series. In this view the hereditary nature of the organism is not seen as a continuum, but is broken down into a series of what Mendel and his immediate followers called 'unit' characters. (We should nowadays think of them rather as unit development functions or biochemical reactions; a character such as the colour of the eyes is a complex affair to which many more elementary biochemical reactions may contribute, but this refinement does not alter the essentially atomistic nature of the analysis which Mendel inaugurated.)

Mendel applied the atomistic outlook not only to the analysis of the organism into a series of unit characters or unit developmental functions, but also to the hereditary factors which control the potentialities for these characters. He suggested that, with respect to any given character, an organism at fertilization receives one factor from its mother and another from its father. The potentiality which the organism possesses for developing the character in question will be determined by the combined action of these two factors. The two factors may be either exactly the same or slightly different from one another. If they are somewhat different, the potentiality which the newly-conceived organism possesses will usually show the influence of both of them, although in some cases one may have a much more powerful effect than the other, in which case, the powerful one is said to be 'dominant' and the less influential one 'recessive'.

Although the two factors which come together in the fertilized egg both influence the potentialities of that egg, the factors themselves do not mingle and blend with one another, but remain quite separate and unaffected by each other's presence. Suppose, for instance, that organism receives from its father a factor p_1 and from its mother a factor p_2, both of which influence its potentiality P. The developmental process which goes on as the egg develops will be influenced both by p_1 and p_2; an animal having the constitution p_1p_2 may in fact be intermediate between animals which have p_1p_1 or p_2p_2. But when the p_1p_2 animal grows up and in its turn becomes a parent, it will pass on to its offspring either the factor p_1 or the factor p_2—in fact half of them will get the first, and half the second—and the p_1 and p_2 factors which are passed on in this way are quite unchanged. The p_1 will still be p_1, and will have none of the p_2 character mingled in with it. The factors p_1 and p_2 have behaved in an atomistic fashion, each retaining its own character, not like fluids which have mingled together into a blend.

From the simple hypothesis which Mendel put forward it is easy to work out how a given factor should behave as it is transmitted from one generation to another. These deductions were immediately verified experimentally. In fact Mendel himself provided a solid basis of fact which entirely supported his theory, although it seems probable that he had, by one of the great imaginative feats on which scientific progress depends, thought of the theory before he had carried out the experiments to support it.

Mendel's work was neglected by biologists in general until about 1900, but in the comparatively short time since then, genetics has grown into perhaps the most highly developed of all the branches of biology. The greater part of this development has relied on one single method of investigation. This makes it comparatively easy to explain and to understand, and it is worth looking at it rather more closely as a remarkable example of the power of an atomistic approach.

The fundamental advances of genetics have come almost entirely from studies which have followed the hereditary transmission of two or more kinds of factors simultaneously. Suppose an animal at fertilization receives p_1 and q_1 from its first parent and p_2 and q_2 from its second parent. What will happen when this animal itself mates and has offspring? Will the p_1 and q_1, which came from one parent, tend to be transmitted together to the next generation; or shall we find that p_1 is just as likely to be accompanied by q_2 as by q_1? Actually, when such matters are looked into, it is found that the hereditary factors in a given organism fall into groups, such that all the members of a group have a tendency to remain together in these circumstances, while members of different groups do not. In the technical, but easily comprehensible, vocabulary of genetics, we say that there is free recombination of factors between the groups, but linkage of factors belonging to the same group. Even between two factors belonging to the same group the linkage is not absolute. If p_1 and q_1 came into an animal together, and belong to the same linkage group, they will tend to be transmitted together, but there will be a proportion of cases in which they become separated so that p_1 is transmitted combined with q_2. It is the exact quantitative estimation of how often this recombination happens that is the major clue to the investigation of genetics.

The first simple rule that emerged deals with the recombination that may occur between three linked factors, let us say, p, q and r. By making suitable crosses and counting the numbers of different types of offspring we can estimate the amount of recombination between each of the three pairs p and q, q and r, and p and r. It was found that these three recombination factors are always of such a kind that one of them is equal to the sum of the other two. For instance, if p and q recombined in 5 per cent of the cases and q and r in 10 per cent, then p and r would recombine in 15 per cent.

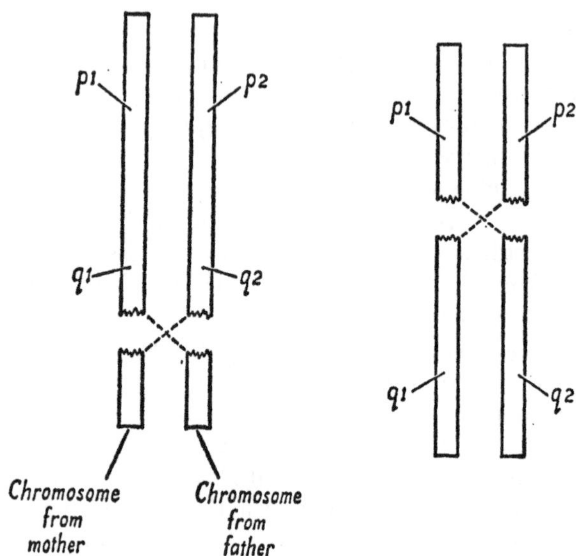

FIG. 2

Recombination of genes. The chromosomes coming from the mother and father sometimes break at equivalent places along their length, and the parts recombine in the wrong way. If the break happens to lie between two genes p and q, as in the right-hand diagram, the maternal and paternal genes will become recombined, so that p_1 now goes with q_2 instead of q_1, and so on.

It was this fact which stimulated Morgan to suggest that the hereditary factor could really be considered as a material particle. The 'additive theorem of recombination fractions', which was explained in the last paragraph, is exactly what one would expect if the factors were material bodies located at different points along a straight object such as a stick, and if their recombination depended on the stick being broken between them. If the position at which the break occurs is purely at random, then clearly the proportion of cases in which it occurs between two factors will depend on the distance between them. The farther apart the factors are the more likely the stick is to break in such a way as to separate them. Now it was already known that there are rod-like bodies occurring in the cell and, moreover, they occur in pairs just as the hereditary factors do. These bodies lie in the nucleus and, because they stain darkly with several types of dyes, are known as 'chromosomes', or coloured bodies.

Morgan suggested that the hereditary factors, which were introduced by Mendel as hypothetical abstract entities to explain the proportion of offspring in various crosses, were simply material structures attached to, or making up, the chromosomes. It was soon found that the number of different linked groups of hereditary units is exactly the same as that of the number of chromosome pairs. Moreover, when one chromosome pair behaves, for some reason, in an abnormal manner and can be seen to do so by microscopical examination of the cells, then it is found that certain hereditary units also behave in a corresponding manner. The evidence that the factors lie on the chromosomes is therefore very complete. Since the time when Morgan demonstrated this it has become usual to refer to them, not by the abstract term 'factor', but by the name 'gene', which implies that they are actual material entities.

As soon as the 'factors' were regarded in this way, genetics made extremely rapid progress. One could, in the first place, try to find the size of the gene. The way to do this is to study genes which are very closely linked together, that is to say, which lie so close side by side that breakages very rarely occur between them. Eventually one might hope to find cases of two genes which lie absolutely next door to one another. Again, one can examine the chromosomes themselves with the microscope. They are, as was said, rod-like bodies, but they do not look exactly the same all the way along their length. Sometimes they show largish portions which stain differently to the rest, sometimes they show a series of lumps, the so-called chromomeres, and in a few cases they show an elaborate pattern of alternating dark and light regions. Finally, one can apply chemical methods to discover the nature of the substances from which the chromosomes and therefore the genes are built, and here an extremely interesting point emerged.

Chromosomes are composed of combination between proteins and a substance known as deoxyribonucleic acid (or DNA for short; there is another variety of nucleic acid, known as ribonucleic acid or RNA, but this plays only a minor role in the make-up of the chromosomes). Now both proteins and DNA are linear structures made up of a large number of small units joined end-to-end. The building blocks of which the proteins are composed are known as amino acids, while those of DNA are nucleotides. In both cases the building blocks are assembled one after another into long chains. Moreover, the length of each link in the amino acid chain of a protein is the same as, or a simple multiple of, the length of each nucleotide link in a DNA chain. Thus, the two chains would fit nicely if they lay side by side. Moreover, this chain-like structure is just the same kind of thing, on a smaller scale, as the linkage of genes to one another in a linear sequence, and as the structure of the chromosomes which, as we have seen, are long rod-like or thread-like bodies with different regions along their length.

These points were all discovered during the first forty years of this century. This is nowadays often considered to be the classical period of genetics. During this time geneticists studied heredity in what one might roughly call ordinary animals and plants. Since then, as we shall see in a moment, new types of organism have been studied genetically, and these have opened some new doors into the nature of the genetic material. In order to bring out the nature of these new advances, it may be as well to look back to the picture as it was at the end of the classical period. In the last few pages of a general account of genetics which I wrote at that time,* I summarized our knowledge of the gene as follows:

'One can start from the fundamental fact that the chromosome is an elongated structure which, whenever we can analyse it, has differences arranged in a linear order along it; these differences can be detected by linkage studies, chromosome structures, etc. The units, between which differences are noted, may be of different sizes according to the different methods of investigation: there are, in roughly descending order, inert or precociously condensing regions, large chromomeres, ultimate chromomeres or salivary gland chromomeres, the units of crossover and X-ray breakage. One might symbolically represent the chromosome thus

abcd'e'f'g'hijklMNOPQRSTU'V'W'

where there are differences on three scales, between the capitals and lower-case letters; normal, underlined and dashed letters; and finally the letters themselves. The smallest units of this scheme, symbolized by the individual letters, are the units of crossing over and X-ray breakage, and probably measure, as we have seen, about a 100 millimicrons in length.

'If we view the chromosome as it were through the other end of the telescope, attempting to build it up from chemical units, we arrive at a somewhat similar scheme of a linear order of units of different orders of magnitude. The ultimate units now are the links in a polypeptide (i.e. amino acid) chain, with a length of only 0·334 millimicrons. Exactly what the larger units are is more doubtful, but we have a range of possibilities; . . . it is, then, possible to conceive of the chromosome as a linear array of units, the units themselves forming a hierarchy all the way from heterochromatic and euchromatic regions, some tens of thousands of millimicrons long, to polypeptide links only a few tenths of a millimicrons long.'

As this makes clear, the classical period of genetics finished with a large gap between the smallest units which could be detected by genetic analysis of the linkage and recombination of genes, and the chemical sequence of units which are built up into protein and nucleic acid. The great advances made in the post-classical period of genetics are, firstly to fill up this gap, and secondly, to suggest that it is the sequence of nucleotides in the DNA, rather than the sequence of amino acids in the protein, that is the essential feature in determining the nature of the genes.

The basis of the first advance is a simple technical one.* Extremely ingenious methods have been worked out for carrying out genetic studies on organisms such as bacteria and viruses. A bacterium, and still more a virus particle, is, of course, enormously smaller than the organisms with which classical genetics had worked, such as mice, fruit flies and maize plants. Once the necessary technical tricks have been developed, these tiny organisms can be handled in millions, whereas it was a major effort to breed more than a few thousand offspring in the old days. Now the larger number of individuals available for study means that one can detect recombinations that occur only exceedingly seldom; and this makes it possible to find out something about points on a chromosome which lie extremely close together and thus recombine only in a minute proportion of cases. By now the most refined analyses, using a virus which is so minute that it lives as it were as a parasite on bacteria, have detected recombinations that occur so seldom that they indicate that the ultimate units which recombine cannot be much larger than the unit amino-acids or nucleotides which build up the protein or nucleic acid chains. The genetic analysis, in fact, has been pushed right down to the point where it is talking about entities of about the same size as those which engage the attention of the chemists. The gap which the classical geneticists had to leave between these two methods of investigation has thus been bridged.

In the time of classical genetics, people who tried to conceive the gene in chemical terms tended to picture it as essentially composed of proteins, which themselves consist of sequences of amino acids. In the last twenty years several lines of evidence—which are perhaps not quite as convincing as they are often thought to be—have persuaded most biologists that it is the DNA rather than the protein which plays the essential role in the chromosome. If this is the case, we have to think of the gene as essentially made up of a sequence of nucleotides.

In a way this is fortunate, since nucleic acids are simpler structures than proteins. There are, for instance, only four main different kinds of nucleotides, whereas there are twenty different kinds of amino acids. Watson and Crick, basing themselves on

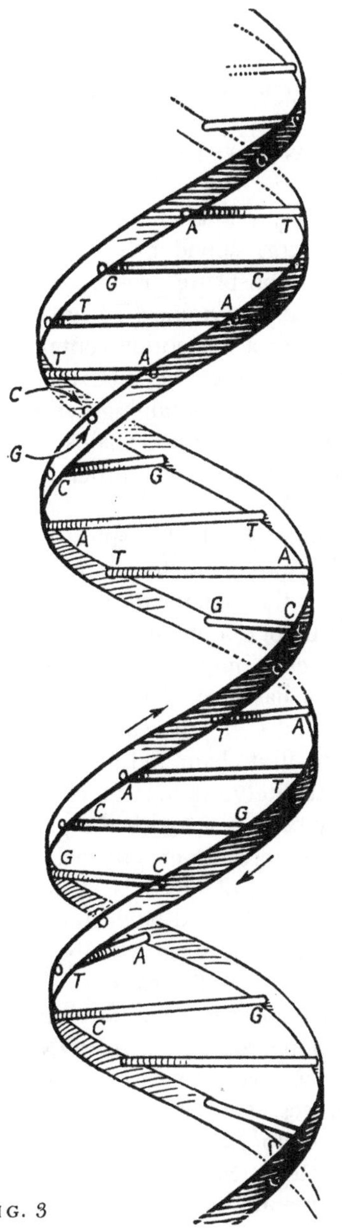

FIG. 3

A diagram of the structure of DNA. It is made up of two threads, which are twisted in a spiral round one another. Along each thread there is a sequence of nucleotides containing one or other of the four bases indicated by the letters A, C, G or T. The bases on one thread are attached to those of the other thread, as indicated by the cross-bars. The rule is that an A can only be attached to a T, and a C to a G. Thus in the (purely diagrammatic) section of DNA shown, the sequence, reading from the top downwards, is:

TCAAGCATG etc. on one thread

AGTTCGTAC etc. on the corresponding thread.

the work of a large number of biochemists and X-ray crystallographers, proposed a few years ago a scheme for the construction of nucleic acid, which is now generally accepted as providing at least the major outlines of the way in which it is built. They suggested that DNA molecules consist of not one linear chain of nucleotides, but of two linear chains which are coiled round one another in a spiral. At any point on the spiral the nucleotides in the two chains have to correspond if they are to fit together. Each nucleotide contains a phosphate group, a sugar, and one of the four bases, adenine (A), thymine (T), guanine (G) or cytosine (C). The rule is that a nucleotide containing adenine can only fit with one containing thymine, and similarly a guanine nucleotide can only fit with one containing cytosine. Thus, if the order of nucleotides in one of the two chains is, for instance, TGCATTC, this automatically fixes the order in the other chain, which must be ACGTAAG.

It is this order of nucleotides along the double spiral which we have to think of as determining the character of the hereditary material at that region. What the genetic analysis shows is that hereditary differences may be of about the same size as these nucleotide links. For instance, it might be that in the sequence given above the third link has been changed, so that instead of having C in one chain and G in the corresponding chain, we now have perhaps T in the first chain and must therefore have A in the second. Unfortunately, as yet the chemical methods of analysis have not fully caught up with the genetic ones, and it is not yet possible to determine exactly what the sequence of nucleotides in a given piece of DNA is. Thus, although the genetic analysis shows that we must be dealing with changes of this kind, we cannot in any given case actually verify the detailed nature of the changes by chemical means.

We can carry the story somewhat further if we now turn to consider the proteins. As well as being combined with DNA in the chromosomes, proteins make up a great part of the whole body of the cell. The chemical processes occurring in cells are nearly all carried out by the aid of biological catalysts known as 'enzymes', and these are in fact proteins. Since the genes in the chromosomes control the developmental processes, this must mean that they specify the detailed nature of the proteins that will be formed.

Although proteins are more complicated molecules than nucleic acids, it has in a few cases been possible to analyse in detail the sequence of amino acids out of which they are built. One

Valine	*Valine*
Leucine	*Leucine*
Leucine	*Leucine*
Threonine	*Threonine*
Proline	*Proline*
Glutamic acid	*Valine*
Glutamic acid	*Glutamic acid*
Lysine	*Lysine*

FIG. 4

Part of the sequence of amino-acids which make up the protein haemoglobin. On the left is the normal sequence; on the right is the sequence in the abnormal 'sickle-cell' haemoglobin, in which a valine had been substituted for one of the glutamic acids.

case in which this has been done is for haemoglobin, the red protein in our blood cells. Now we know a few genes in man which affect the nature of the haemoglobin. There is, for instance, the very interesting 'sickle-cell' gene, which changes haemoglobin in a way which can be easily detected when the blood cells are treated in a certain way; cells from a person showing the sickle-cell character shrink to a sickle shape, while normal blood cells remain round. The sickle character is perhaps somewhat harmful under normal circumstances, but it seems to confer some degree of resistance against malaria, which is, of course, a disease of the blood cells. Presumably as a consequence of this, one finds that the gene is by no means rare in certain human populations in which malaria has been a major hazard, as in many tropical regions of Africa and Asia. When sickle cell haemoglobin was analysed chemically it was found that it differed from normal haemoglobin only by a change of one amino acid in the sequence. There are, as was said above, twenty different amino acids arranged one after another in a protein. In sickle cell haemoglobin at one place in this chain an amino acid known as valine is inserted in the place that should be occupied by another amino acid known as glutamic acid. In another hereditary variety of haemoglobin, known as C haemoglobin, another amino-acid, lysine, has been substituted for a glutamic acid; oddly enough, it is exactly the same glutamic acid in this case also, a phenomenon for which there is as yet no very good explanation. The point that emerges from this is that different genes determine different sequences of amino acids in the proteins, and that a change in a gene can change a protein merely by substituting one single amino acid for another at a particular point in the chain.

This faces us with a pretty puzzle. We have the hereditary material consisting essentially of sequences of four different types of nucleotides; we might regard it as a language with only four letters, A, T, C and G. In some way the order in which these four letters are placed has got to determine sequences of twenty different amino acids. How can this be done? It is clear that it could not operate if one nucleotide corresponded to one amino acid; since there are only four types of nucleotide they could only specify four types of amino acid. We have to suppose that each amino acid is specified by a 'word' of two or more letters of the four-letter alphabet of the nucleotides. Two-letter words still won't do the trick; they could only specify four multiplied by four = sixteen different amino acids. Three-letter words, at first sight, seem capable of doing too much. There are $4 \times 4 \times 4 = 64$ of them, and there are only 20 amino acids whose position we require to fix. But there is another point to be thought of. Consider again the sequence TGCATTC, which was mentioned a few pages ago. If this is to be interpreted as a series of three-letter words, each of which corresponds to an amino-acid, we have to know where in the sequence to find an initial letter with which to begin. Is the sequence to be interpreted as TGC ATT C, or perhaps as T GCA TTC? There are no gaps along the sequence of nucleotides to break it up into three-letter groups, so some other way has to be found of doing so. One way, suggested by Brenner, would be this. We might suppose that some combinations of letters make a word which corresponds to an amino-acid, while other combinations are simply nonsense and do not correspond to anything. Then a sequence would be unambiguous if it is made up of three-letter blocks which are sense while the groups of three which overlap between these blocks are nonsense. For instance, if TGC and ATT are both words corresponding to a particular amino acid, while GCA and CAT, etc. correspond to nothing, then there would be no possibility of misinterpretation. If one tries to work out systems of this kind, it turns out that with a four-letter alphabet and three-letter words one can get not 64 but exactly 20 words that have a meaning; and 20 is the number of amino-acids now known to occur in natural proteins. It seems almost too good to be true, but it has the great scientific merit that the theory has led to a conclusion which can be investigated. If somebody now discovers a twenty-first amino-acid in a protein we shall either have to abandon the theory or at least modify it.

Of course we still have to keep in mind the possibility that the relation between the sequences of DNA nucleotides and protein amino-acids may not really be as direct as these theories suggest. What we know is that both nucleic acids and proteins consist of chains of units, that the units in the two cases are very similar in size, and that in both substances the alteration of a single unit is important. But after all the same is true of English and French words; and it does *not* follow that if you alter one letter in an English word, the French word by which it is translated will also be changed in only one letter. It may possibly be that the reason why biological organisms have gone in for linear molecules like nucleic acids and proteins is to be found, not in the possibility of letter-by-letter correspondence between them, but merely in the fact that they are the simplest form that a very complex but very specific substance could take. If you have to arrange a lot of different things into a definite pattern, it is in a real sense easier to do so by putting them one after the other in a precise order than by disposing them in particular places on a two-dimensional plane or even in a three-dimensional space. Possibly we are hoping for too much in seeking for a letter-to-letter correspondence; but it is certainly worth going on looking for it until we either find it or it is proved not to exist.

Even if the pessimism of the last paragraph turned out to be justified, these recent developments of genetical theory are a wonderful example of the power of an atomistic approach. By considering genes as discrete, separate material bodies, and then their sub-structure as a sequence of small molecular building blocks, we have been led right down into the intimate chemical details of the materials which determine the characters that living things have. This is a scientific triumph as striking as the analysis of the ordinary material objects of our environment, the chairs we sit on and the tables we write at, into electrons and protons. But even in this field of genetics atomicity is not the whole story. There are other aspects of the phenomena we are examining which require something more like a continuum approach. These facts, which make one doubt whether the atomistic theory is quite adequate by itself, do not all hang together into a quite coherent story, so that in expounding them it will be necessary to jump about to some extent from one type of experiment to another.

First of all, it is necessary to point out that in the workings of the cell itself the connection between the nucleotide sequences in the DNA and the amino acid chains in the protein is rarely, if ever, a quite direct one If the genes in the chromosomes directly controlled the formation of proteins, then we should expect to find the proteins in the cell being produced inside the nucleus. A good way of investigating where protein is formed is to provide a cell with amino acids which contain an atom that is radioactive. If protein is being synthesized in the cell the radioactive amino-acid will be built into it. One can then kill the cell and cut it in thin slices, and place over each slice a film of photographic material. The radioactive atoms produce ionizations, which act on the photographic emulsions just as light acts in a camera. After a suitable length of time—and the 'exposure' for this purpose may amount to a few days or even a few weeks rather than a fraction of a second—the film can be developed; and wherever it becomes blackened we know that radioactive material has there been incorporated into the substance of the cell. There are, as might be expected, various difficulties in interpreting such experiments, but when used with care they give a good indication of the sites in the cell at which new protein is being formed.

What we find* is that in normal cells the most active place for the formation of protein is not in the nucleus, but outside it in the cytoplasm. In very early embryonic cells, it is true, the nucleus seems to be the position of most active protein synthesis. But these are cells which are only just beginning to develop from the simple condition they had as parts of the egg, and they probably still lack much of their normal cellular machinery. It is in cells which have already differentiated further than this, and become liver cells, muscle cells, nerve cells and so on, that we should expect the genes to be functioning at full speed. Yet in rather more mature cells of this kind it is not in the nucleus that the proteins are being produced most rapidly, but on the contrary in the cytoplasm.

In fact the protein seems to be formed mainly in connection with a series of tubules and flattened vesicles which fills up most of the cytoplasm in actively synthesizing cells. This system at present has two names, both rather formidable ones, the 'ergastoplasm' or the 'endoplasmic reticulum'. When cells are killed and sliced extremely thinly, and the sections examined with an electron microscope, the ergastoplasm has an appearance which reminds one of an air view of a housing layout in the suburbs. There are usually what look like long gently curving avenues with a series of little dark granules (corresponding to houses) on either side. These 'avenues' seem in most cases to be sections through flattened vesicles. The structure is really more like a pile of deflated rubber balloons loosely pushed together. If the cell is smashed up it is possible to separate from the soup a collection of the little black granules. It seems to be in these, which are known as microsomal particles, that the proteins are most rapidly formed in fully developed cells. Now these particles themselves do not contain any DNA; instead they have the related compound known as RNA. For these and other similar reasons it is rather generally agreed nowadays that RNA is involved somewhere in the process of the formation of proteins, coming between DNA and the protein itself. This is a first complication in the simple story relating sequences of nucleotides to sequences of amino acids.

If we go back to the genetical analysis of the hereditary material we shall find another kind of complicating circumstance. It is not sufficient to consider only the relation between an amino acid and the three (or whatever number it may be) nucleotides which correspond to it in the DNA sequence. There are longer stretches of DNA than this which behave in some ways as units, and refuse to be completely atomized. In fact there is probably a whole hierarchy of such larger units, corresponding rather to the hierarchy of underlined, lower case, and so on letters on page 37.

Probably the smallest of such more extended unit is a kind which puts in an appearance when we consider the changes that genes may undergo. The hereditary material is, of course, extremely stable, so that it can be transmitted through many generations. Its stability is, however, not absolute. If it were, no new hereditary variations could arise and there could have been no evolution. In point of fact the genetic material occasionally changes. The process of change, which is known as 'mutation', may occur spontaneously, but the frequency with which it happens can be increased by various agents, such as X-rays or other ionizing radiations and certain chemicals. Probably the change in most cases consists in the substitution of one nucleotide for another in the DNA chain; for instance, a nucleotide containing G may take the place of one containing A. When a very small organism such as a virus is used, and X-rays are employed to stimulate the changing of the genes, enormous numbers of mutations can be collected. It has been found that they do not occur completely haphazardly along the length of the DNA chain; there are certain regions, known in laboratory jargon as 'hot spots', which give many more mutations than others. These hot spots seem to have the length of a dozen or two nucleotides; at any rate, they are somewhat too large to correspond to a single amino acid. The precise interpretation of the phenomena is still somewhat obscure, but here we seem to have evidence of regions which are rather larger than those corresponding to a single amino acid, and which yet have a certain unity by sharing in the property of being a hot spot.

The next largest unit is one known as a cistron. It reveals its presence in this way. Consider a cell which, like most cells, contains a number of *pairs* of chromosomes. Suppose one of these pairs consists of the two chromosomes C_1 and C_2. Now think of two abnormal genes g_1 and g_2 which are carried on these two chromosomes. Clearly there are two possible situations for the cell; either g_1 lies on one of the chromosomes, say C_1, and g_2 lies on the other C_2; or both the abnormal genes lie on one of the two chromosomes, say C_1, and the other chromosome C_2 is perfectly normal. Breaking into dog Latin, as they do at times, biologists call the first situation the trans-configuration and the second the cis-configuration.

When we examine a number of examples of these two situations we usually find that cells containing two abnormal genes in the cis configuration are nevertheless perfectly normal. Presumably this normality is brought about by the presence of the other normal, chromosome C_2. When the genes are in the trans-configuration, however, we find some cases in which the cell is normal and others in which it is abnormal. When it is normal we have to conclude that the cell contains one adequate representative of all the functional units it needs. We have to suppose, therefore, that corresponding to the abnormal factor g_1 there is a normal factor G_1, and corresponding to g_2 another normal factor G_2. Thus, normality in the trans situation arises when there are two effective factors, G_1 and G_2. When the cell is abnormal, however, we have to suppose that there is no normal factor present. In that case the two factors g_1 and g_2 must both be abnormal forms of the same functional unit, the normal type of which we might call G. The factors defined in this way

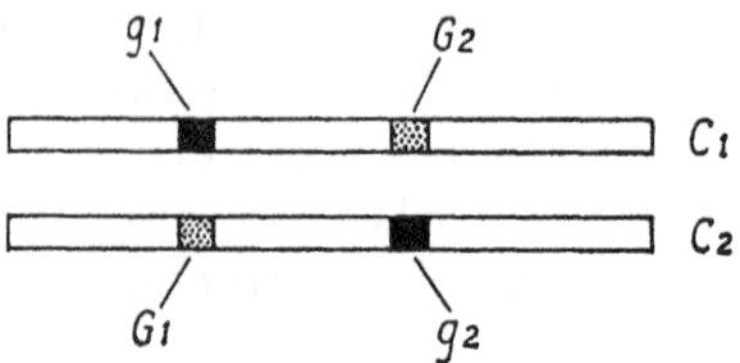

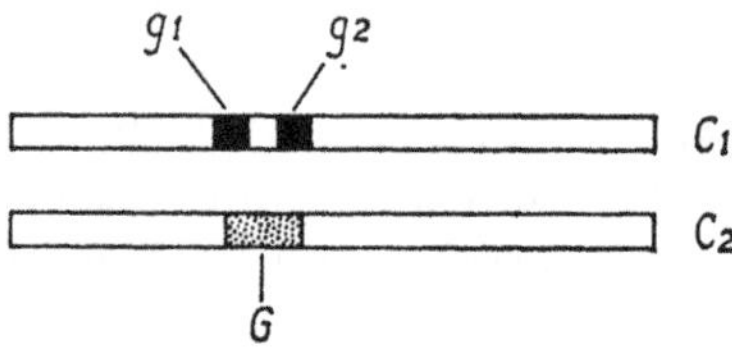

FIG. 5

The cistron. In the upper drawing, there are two abnormal genes, g_1 and g_2, one on each chromosome, that is to say, in the 'trans' arrangement. If the individual with this hereditary constitution appears normal, that indicates that it contains normal genes corresponding to both g_1 and g_2, as indicated by G_1 and G_2; and we conclude that g_1 and g_2 lie some distance apart along the chromosome. In the lower diagram, g_1 and g_2 are in the same chromosome, i.e. in the 'cis' arrangement. The individual is normal because of the normal gene G. But if, with these same factors, g_1 and g_2 were in different chromosomes ('trans') there clearly would be no normal G. The individual would then be abnormal and we could conclude that g_1 and g_2 belong to the same 'cistron'.

(G, G_1 or G_2) are what are referred to as 'cistrons'. As this scheme makes clear, the cistron is a functional unit, and the cell can only be normal if it contains at least one unaltered representative of each of the cistrons involved in its operations. Now these cistrons, which are clearly units of a very important kind, are again much larger—by a factor of say 1,000—than the stretch of DNA which corresponds to a single amino acid.

But the cistrons are still by no means the largest units we have to consider. What, in the light of our new knowledge, corresponds to the old-fashioned gene? At first sight one might think that the answer would be, the cistron. In the classical days of genetics one of the most important characteristics of the gene was that it behaved as a unit of function; that is to say, a gene could from this point of view be defined as the region of a chromosome which was involved in some unitary physiological activity, such as for instance, the production of an enzyme of a specific kind. This is just what the cistron is doing. But, rather surprisingly, it turns out that the cistron, defined in the way which was sketched above, is considerably smaller than the old-fashioned gene. In fact, many chromosome sections, which in classical days were considered to be single genes, have turned out to contain three or four or more cistrons.

For instance, there is a gene in one part of the Drosophila sex chromosome which controls the formation of pigment in the eyes. If an animal contains only mutant (abnormal) forms of this gene, its eyes will be colourless or very pale. In classical days this was regarded as one single unitary 'white-eyed gene'. However, in recent years several people, stimulated by the results obtained with bacteria and viruses, have had the patience to breed enormous numbers of flies containing these genes, and have thus discovered that there is some recombination between them. This makes it possible to arrange the different abnormal forms of the gene in a linear series corresponding to the places where the DNA chain had been altered. It was then found that all the genes which had been altered towards one end (known as the right end) had a slightly different effect in interaction with other genes than those which had been altered at the other, left, end. From such studies it has appeared that the 'white-eye gene' really consists of at least four spatially separate regions, which differ from one another to some extent in their activity, although they are all concerned with the formation of pigment in the eye. Again, there are other genes known, which in the old days would have been considered single and unified entities, but which have more recently been shown to give rise to mutant forms which 'complement' one another, that is, the organism which contains one mutant in one chromosome and the other mutant in the other chromosome appears quite normal.

This means that some of the old genes really consist of several cistrons, and are therefore not completely unified entities. Nevertheless, the whole group of cistrons which make up a 'gene' must have some degree of interconnectedness. The group was originally considered a single gene just because an alteration of any part of it produced much the same kind of effect on the animal bearing it—as any alteration in the old 'white gene' tended to produce a white eye, for example. The nature of the inter-connections between the cistrons which form a set of this kind is still highly obscure. In a very few cases, which are found particularly in bacteria, the group of cistrons seems to operate on a sequence of chemical reactions. For instance, suppose that substance A is converted into B, that into C, and that into D and so on; a few examples are known of a set of cistrons arranged in exactly corresponding order, with a cistron C_a affecting the formation of A, followed by a cistron C_b affecting the production of B, followed by cistron C_c affecting C, and so on. This would suggest that the cistrons are arranged along the chromosome like a set of machines along a production line in a factory, and that the chemicals they affect are passed on from one to the next when ready. However, this does not seem to be a general rule, and some cases are known which cannot possibly be explained in this way. For instance, in certain fungi the cells often contain two separate nuclei instead of only one. If there is a set of two complementing cistrons C_a and C_b, we can have an abnormal version of C_a in one nucleus and an abnormal form of C_b in the other; and it is found that in some cases complementation will still occur so that the cell is normal.* Here the unaltered cistron in one nucleus is co-operating with the unaltered cistron in the other nucleus, even though they lie a long distance apart and are separated by a region of cytoplasm. Although the mechanism of co-operation in such cases of complementation is still unknown, it is nevertheless clear that the cistrons fall into larger groups or sets which are functionally closely related.

At least one case is known (bithorax in Drosophila*) in

which there is a certain parallelism between the position within the 'gene' of the altered cistrons and the region of the body which this alteration renders abnormal. This situation is even more difficult to understand than those that have been mentioned above. It serves to remind us how extremely little we know about how the genes in the chromosomes control such essential features of biological organization as the appearance of pattern and shape. The studies which have led to the atomistic theory which relates nucleotides sequences to amino-acid sequences have been concerned primarily with the effect of genes in controlling the production of definite chemical substances. Although these substances are proteins and therefore to a chemist seem very complicated, they are simplicity itself compared to biological entities such as a leg or a vertebral column. In such major portions of an animal's body we are confronted with something which is exceedingly complex in detail but which has a relatively simple overall shape or morphology. This has tempted some geneticists, such as Pontecorvo, to suggest that there may be functionally effective 'higher fields' in the genetic material; that is to say, large groups of cistrons which work in some way as units in the determination of the overall morphological patterns of the developing organism.

Whether such higher fields really exist or not, there are certainly some further steps, in the hierarchical organization of the genetic material, larger than the functional sets of cistrons which make up the old-fashioned genes. There are, for instance, the differentiated regions of the chromosome which are large enough to be visibly detected, such as the deeply staining knots known as chromomeres, or the dark bands seen in the giant chromosomes in certain cells of flies. On a still larger scale, there are the sections of chromosome which the microscopists distinguish, on the basis of their reaction to certain stains, as 'heterochromatic', or 'euchromatic'. These larger units of organization were, of course, known in classical days, but the great advances that have taken place since then in our knowledge of the detailed structure of the chromosomes do not render them any the less real.

In fact our newer knowledge does not weaken, but rather re-enforces, the general scheme of the organization of the genetical material which had already emerged some twenty years ago, and which I symbolized by representing a chromosome as a hierarchy of linearly arranged elements of different sizes in the form

abcd′e′f′g′hijklMNOPQRSTU′V′W′

The great triumphs of the atomistic approach in recent years have made it possible to put forward hypotheses about the nature of the basic units—the single letters in the scheme above—which are both more precise, more plausible, and more subject to experimental verification, than any ideas which were current twenty years ago. But we have also to remember the points which call for a mode of thinking which lays its emphasis on organization rather than on atomicity; the facts that are symbolized by saying that the individual letters are grouped into words, and those into sentences, and paragraphs, and chapters.

CHAPTER 3

Development

ANY attempt to analyse living organisms into their simplest components must lead us to a consideration of the hereditary materials, but those materials only specify the potentialities which a new individual inherits from its parents. In order to understand organisms as they actually occur around us we have to discover how those potentialities become translated into realizations. We are thus confronted with the second of the biological time scales, that of development. As far as genetics is concerned, the hereditary material is simply something which is passed on from one generation to the next, but in their activity the genes are essentially agents which operate on developmental processes. As the last chapter made clear, we cannot even discuss their constitution fully without bringing in considerations about how they act. When we turn from the attempt to reduce living things to their simplest terms and attempt to understand them in the full concreteness of their existence, we have, in the first place, to think about developmental processes in much fuller detail.*

A living organism is not just a bag of chemicals each produced by the influence of some particular gene. It has a character which we acknowledge by calling it a living organism. This phase admits that it exhibits the property of organization; but what exactly is organization? It is a rather tricky concept to define, and it is probably sufficient to say here that it implies that if an organized entity is broken up into parts, the full properties of these parts can only be understood by reference to their relations with the other parts of the whole system. For instance, the nature of a man's leg can only be fully understood when it is seen as having a certain relation to the whole body. Or again, we must inevitably omit something important from an account of the lens or the retina of an eye unless we realize the relations between these parts and the eye as a whole functioning unit.

It is perhaps clear on first principles that the development of entities which are organized in this sense must demand something more than a purely atomistic theory. We have then in the study of development rather the opposite situation to that which confronts us in the study of heredity. Whereas the latter has seemed, since Mendel's day, to cry aloud for an atomistic theory, the former seems to demand organismic or non-atomistic theories. But, as we saw in the last chapter, it is necessary to supplement the atomistic theories of heredity by some considerations of the continuum type, and one of the points which we shall have to explore in this chapter is whether, or how far, atomistic notions are valuable as correctives to purely continuum theories of biological organization.

The first major fact to be noted about biological organization is that it is essentially a dynamic affair, involving the lapse of time. The organization of an animal arises gradually as the egg develops into the adult, and to understand the organization we must follow the changes and processes by which it comes into being. The first experimental studies of such questions led to theories which were very far indeed from being atomistic. In fact they were most often completely at the other end of the spectrum of possible types of theory.

For instance, towards the end of the last century, Driesch performed some of the earliest experiments on developing eggs. He took the newly fertilized eggs of sea urchins and cut them in fragments. When these were allowed to develop he found that each fragment seemed capable of producing a complete and normal embryo. Driesch concluded that no mechanistic, let alone atomistic, system could possibly explain such a result, and that it was necessary to suppose that development is controlled by some non-material agency, which he called an entelechy. His theory was in fact a vitalistic one.

We need not, however, pursue it much further, since the actual facts of the situation are not quite as Driesch thought. In practice, when he cut his eggs into fragments, he sliced them always from top to bottom. If he had turned them on the side, and cut them through so as to separate the top half from the bottom half, he would not have found that these fragments gave complete embryos. Since his day, such experiments have been very extensively performed, particularly by a school of Swedish biologists led by Runnström and Hörstadius. They have shown that the upper half of the sea urchin egg has a character quite different from that of the lower half. Normal embryos arise only when the upper, or as it is called 'animal', character is in correct balance with the lower, or 'vegetative', character. This balance can be achieved even in certain abnormal combinations of fragments. For instance, if a small portion of the upper end of the egg (less than half of the whole egg) is combined with a suitably sized small portion from the lower end, the egg has a correct balance between the properties of its different regions, and these interact with one another to control normal development. This brings the system into the range which can be accounted for in terms of organismic theories, and removes the need to postulate any non-material vitalistic principle.

The great importance in development of interaction between different parts of the system was particularly brought home to biologists by some classical experiments by the German embryologist Spemann. Although Spemann's work was perhaps not so revolutionary in its consequence as that of Mendel, his experiments can not unjustly be regarded as the starting point for the growth of modern ideas on development, just as Mendel's were the origin of modern genetics. They were carried out before the Swedish work on sea urchin eggs referred to in the last paragraph, and in fact provided a stimulus for them.

The basic experiment was very simple. It was performed on the eggs of newts. These eggs, like all others, begin their development by becoming divided up into a large number of cells, each of which is considerably smaller than the original egg at the time of fertilization. This packet of moderately small cells becomes arranged as a hollow sphere. One can put the point of the experiment in its simplest possible terms with the aid of a diagram like that in Fig. 6. This shows the egg, as seen from the side as it floats in water. What Spemann showed was that the two quadrants labelled 1 and 2 are at first com-

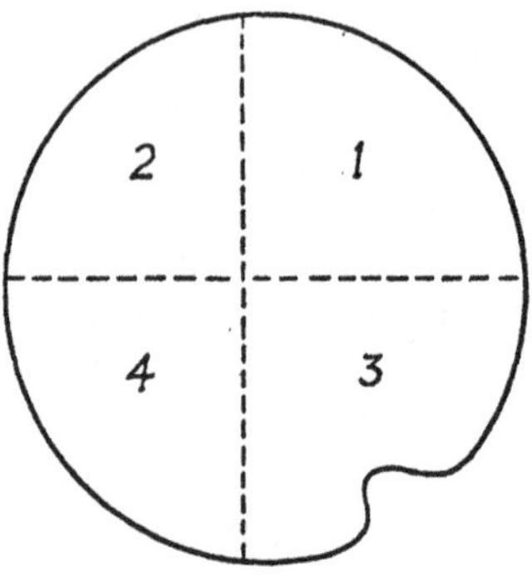

FIG. 6

The newt's egg, seen from the side, and marked off into four quadrants.

pletely similar in their properties, that is to say, in the way they will develop in a variety of circumstances. Later on quadrant 3 is pushed into the interior of the hollow ball of cells and comes to lie underneath quadrant 1. Just after this quadrant 1 develops into the beginning of the nervous system, and Spemann proved that it does so because of some influence on it which is exerted by quadrant 3. In fact, if pieces of quadrant 3 are cut out and placed underneath part of quadrant 2, the region with which they are in contact will also develop into nervous system.

Spemann found, in fact, that the egg has from the beginning some special property in quadrant 3, and that if, as development goes on, quadrant 3 is brought in contact with any part of the upper half of the egg, it will influence that part in such a way as to make it develop into a nervous system. Even if a part of quadrant 3 has been grafted into an abnormal position in the egg, and has thus come in contact with some unusual part of the upper half, the nervous system which it produces is often quite well organized; there may be a fore-brain, with its attached eyes, at one end, and a mid-brain, hind-brain and spinal column arranged in their proper places behind this. Spemann therefore called the region of quadrant 3 the 'organization centre' of the egg. He spoke of small parts of this quadrant, which might be cut out in grafting experiments, as 'organizers'; and he said that they 'induce' the formation of the embryonic organs from the regions of the upper half with which they come in contact.

At one time it began to look as if the process of induction, or at least some aspects of it, could be dealt with by means of a theory of an atomistic type. This would have brought the most fundamental events in embryonic development within the scope of this type of theory. Spemann himself had little inclination towards theories of this kind; in fact, originally his outlook was not very far removed from vitalism. I well remember when I first went to his laboratory in 1931 he firmly believed that organizers could only be effective if they were alive, although his student Marx had just shown that some activity persisted even when they were anaesthetized with a narcotic. At that time I was myself working in this field. I had shown that induction occurs not only in the amphibian embryos with which Spemann worked, but also in those of birds. In the embryo of the chick, for instance, there is also an organization centre, and if little fragments of this are cut out and grafted into a new position in the embryo, they will induce the formation of a new nervous system and other embryonic organs from their surroundings.

In studying the embryos induced in this way in abnormal positions, I found that sometimes they were of a configuration which seemed to demand an explanation by a continuum-type theory.* For example, a grafted piece of organizer may combine with the results of its inducing action in such a way that the two together build up a single more or less complete embryo. Here the activities of the inducing organizer and the reacting material both look as though they were determined by the relation of these parts to each other. But there were also many induced embryos in which this was not so. For instance, if an organization centre is left still attached to the material which would normally form the associated nervous system, and then provided with a second lot of similar material in addition, induction may still take place not only in the normal but also in the extra material, so that one finishes up with two complete nervous systems; that is to say, with much too much to build up a single unified, complete and normal embryo. When this happens, it is impossible to say that the induction is merely the expression of a necessary relation between the organization centre and the rest of the embryo. The organizer seems to be 'inducing' quite regardless of whether more nerve tissue is necessary to complete the system or not. The 'induction' is, in fact, in these cases quite distinct from the 'organization' of the resulting tissues into a unitary embryo. It is difficult to discuss this in Spemann's terminology, in which induction and organization were more or less inextricably tied together, since in his terms the main characteristic of an organization centre is that it induces. I therefore use the word 'individuation' to mean the arrangement of tissues into a unitary system; and my friend Needham later provided me with the word 'evocation' to mean the calling forth of something which otherwise would not have developed, a process which I had previously referred to as 'induction as such'.

If 'induction as such', or 'evocation', can occur quite independently of organization, then there seems no reason why this aspect of development should not be dealt with by some atomistic type theory. For instance, one might consider that it is due simply to the production by the organizer of some active chemical which stimulates the surrounding cells to develop into something they otherwise would not have turned into. But this suggestion was not to Spemann's anti-atomistic taste. I spent the summer of 1931 in his laboratory and that of his pupil Mangold in Berlin. They were rather discouraging about the prospects of attempting to look for an active chemical substance emanating from the organizer. However, as soon as I got back to my own laboratory in Cambridge I started doing such experiments with my own material, the chick embryo. And when the newts started laying eggs in the next spring, Spemann himself, and several of his associates in Germany, also made experiments of this kind. It immediately emerged that both in the amphibia and in the chick, organizers could induce, or as I should say, evocate, even after they had been killed. There was therefore quite certainly nothing vitalistic about the process and indeed it seemed that a theory of atomistic type might be called for.

It is important to remember—as not all authors have done—that the mere production of something new out of the surroundings is not the whole of the process of embryonic induction. It is in fact only that aspect of it which I refer to as evocation. It may be quite reasonable to hope to explain this in terms of the action of some chemical substance given out by the organizer, but, even if we are atomistic with regard to evocation, this does not necessarily mean that hypotheses of the same kind are adequate to explain the whole process of development. A dead lump of material, for instance, may cause its surroundings to develop into a well-shaped complete embryo. Now, since the dead lump can obviously form no part of the embryo itself, there is nothing to prevent us accounting for this evocation of the embryo in atomistic terms. But the embryo itself will exhibit a thorough organization, with all its different regions, such as the brain, the spinal cord, the kidneys, the muscles and so on, having their normal shapes and being arranged in the normal relation to one another. The inter-relations between the parts of the organized embryo are obviously something quite different to the relation between the dead lump and the embryo. In fact, it would seem rather obvious that to account for this individuation of the developing tissues into a normal, well organized whole, some sort of continuum theory would be required. It was probably because some people failed to realize this, that the application of atomistic theories to one aspect of the process of induction was sometimes taken to suggest that the whole process of embryonic development could be dealt with in the same way.

As a matter of fact, even if one leaves the complex process of individuation on one side and confines one's attention to evocation, the atomistic theories have been less successful than could have been hoped. It is true that for a year or two it looked as though it would be relatively easy to discover the nature of the chemical substance given out by a organizer. The main difficulty seemed to be merely that the amount of material in an organizer is extremely small, so that the work would have to be done on a microscale. However, this was too optimistic. Within a very few years Needham, Brachet and I found that we could get evocation with chemical substances, such as the dye methylene blue, which quite obviously and certainly do not occur in the normal embryo. This must mean that the cells which react to the organizer are in some way very ready to develop into the various tissues of the embryo, and that they can be stimulated to do so, not only by the organizer itself, but also by several other things, some of which may, perhaps, operate in ways quite different to that of the natural substance. This makes it extremely difficult to discover exactly what substance the organizer actually gives off in the course of normal development. The problem is, in fact, technically so difficult that it is still being very actively studied now, a quarter of a century later. At the time when the snag first arose it looked almost insuperable. We three, who had first come across it, all decided that, although it was clear that some process of an atomistic type is involved in evocation, we should do better for the time being to turn our attention to something else, rather than trying to identify the actual chemical compound which is effective.

One of the most obvious places to turn is to consider the other half of the evocation reaction. An organizer gives off something which stimulates the cells surrounding it to develop in definite ways. What enables these cells to respond to the stimulus? Now cells normally only develop in ways for which their hereditary constitution has given them the potentiality. The study of how cells can react to an organizer therefore leads us back to the genetic factors and their activity. But we are now approaching the hereditary material from an examination of developing embryos. What we see in these embryos are things like muscle cells, nerve cells, kidney cells and so on. These are much more complex than the single proteins like haemoglobin which we were talking about in the last chapter. A cell contains very many different types of protein, and must involve the activity, not just of single genes, but of large numbers of genes. As we shall see, the genes in these large groups do not act quite independently of one another; their activities interlock to form systems, so that we shall find ourselves needing hypotheses which are not purely of an atomistic type.

An egg begins its development with a set of genes (or if you like a set of pairs of genes) in its nucleus. The egg then becomes divided up into a larger and larger number of cells. At first, each of these cells certainly gets all the genes which were present in the egg immediately after fertilization. It is possible that in later stages of development some of the genes are 'lost', or reduced to complete inactivity, in some of the cells. This is by no means certain, and, if it occurs, it seems likely to be a rather secondary matter and not the fundamental explanation of how differentiation happens. How then are we to explain the facts, firstly, that the character of the egg changes as time goes on, and secondly, that the various parts of it become obviously distinct and different from one another: one part, for instance, turning gradually into a brain and another part into a liver, and so on?

It has been clear for a long time that, in general terms, the explanation is that the egg, when fertilized, is not simply a featureless bag of cytoplasm containing a nucleus, but has a certain degree of structure. That is to say, it contains a certain number of different parts arranged in a definite relation to one another. As the egg becomes divided up into smaller cells, some nuclei will find themselves in the cells derived from one part, and other nuclei in cells derived from a different part. We must suppose that these initial differences in the parts of the egg stimulate into activity different groups of genes (or perhaps one should really say different groups of cistrons; but this modern refinement of terminology is hardly necessary in this connection). For instance, in the newt's eggs after fertilization, one can see a special part, which is grey in colour and is referred to as the grey crescent. It is the cells that are formed out of this part of the egg that later form the 3rd quadrant referred to above, and act as the organization centre. The character of the grey crescent material must be such that in these cells certain hereditary potentialities begin to be realized. In the upper part the egg which forms the quadrants 1 and 2, a different set of genes will be brought into operation. When quadrant 3 moves inside the egg and comes in contact with quadrant 1, and induces the nervous system from it, then this inducing action must involve changing the properties of the cytoplasm of the quadrant 1 cells in such a way as to bring into activity the genes concerned with synthesizing the proteins of nerve cells.

In a few cases one can see with the microscope definite evidence of the activity of particular sections of the hereditary material in particular types of cells. Unfortunately, this is not possible in eggs in which one can see the localized cytoplasmic differences (with the possible exception of one or two special cases in which large fragments of chromosome become lost in certain of the early embryonic cells). The most detailed observation of differences of the activity of different genes can be made in certain special cells of flies. In these cells the chromosomes are very large, in fact, probably each so-called chromosome really consists of several identical chromosomes lying side by side in register. The result is that one sees long thread-like structures marked with a certain number of cross striations. In different types of cells some of the cross striations may be more fully-developed than in others. Again, in some stages of the life history one can see that in a particular type of cell a certain cross striation may show signs of activity, for instance, by swelling or giving off small droplets of material. Quite recently the crucial experiment has been made of changing the surroundings of chromosomes of this kind (by putting the nucleus containing them into the mashed-up cytoplasm of eggs) and showing that this affects the pattern of the activity of the cross striations.*

It is one of the major features of animals—so obvious that its importance is often overlooked—that they are built up of a limited number of rather sharply distinct types of cells. We find nerve cells, muscle cells, kidney cells, liver cells and so on, and each of these major types may have a certain number of minor variations, but we find very few cells that can be considered as really intermediate between the main types. Animals do not shade off gradually from type of structure to another, but consist of definite organs sharply distinct from each other. Development does not produce a continuous spectrum of possibilities, but results in a discontinuous set of different types of cell. Now each type of cell is complex and involves the activity of many genes. The gene activities must therefore be organized into mutually exclusive systems.

These systems are, of course, dynamic affairs.* Embryonic cells, such as those of the upper half of a newt's egg, do not simply sit about retaining constantly the potentiality of developing either into nervous tissue or into skin and so on. They are changing their character all the time, from fertilization onwards. It is only after a certain time has elapsed since fertilization, when these progressive changes have gone some distance, that the cells become ready to respond to an organizer which may switch them into developing as a nervous system or not. If they are left too long without being acted on by an organizer, their readiness to respond will disappear again. There is only a certain phase in their progressive changes in which they are, as we say, 'competent' to react.† And when a system of gene activities has been stimulated, either by the initial nature of one region of the egg, or by the cytoplasm which has been modified by an organizer, the cell does not suddenly and directly turn into a fully developed nerve, or muscle, or whatever it may be. It is only started off on a long course of development, in which one stage will gradually succeed another until the adult condition is reached. We may picture this by supposing that at each stage a certain group of genes is active and synthesies certain proteins; the presence of these proteins will modify the cytoplasm of the cell and this may bring into action some other set of genes, forming new proteins, thus changing the cytoplasm again, and stimulating still further genes, and so on.

The fact that development finishes up by producing a set of sharply different end-results, with few intermediates between them, means that these progressively changing gene activities interact with one another in such a way that only certain paths of development are possible. It appears that, in a cell in which the genes concerned with making muscle proteins are operating reasonably fast, those concerned with making nerve proteins cannot also be vigorously active at the same time. If the cell

† It is the absence of these progressive changes, and thus of the restricted periods of competence, which is the weakest point in the many analogies people have been tempted to draw between the processes of embryonic differentiation and various phenomena which have been found in micro-organisms.

starts making a fair amount of one type of protein, then other alternative activities are shut off.

There seems to be no generally recognized word to indicate a path of change which is determined by the initial conditions of a system and which once entered upon cannot be abandoned. I have suggested for this idea the word 'creode' from the two Greek words *χρη* necessity and *ἁδος* a path. We can say then that the hereditary materials with which an organism begins life define for it a branching set of creodes. Different parts of the egg will move along one or other of these creodes, so that they will, after a long process of progressive changes, finish up as one or other of a number of different end-results, as it might be heart, muscle, nerve, kidney and so on.

Of course, development is by no means completely inflexible. The differentiating cells have a tendency to reach their definite end-result, but if powerful external influences impinge on them they may be to some extent diverted, and finish up in a somewhat abnormal condition. A path of development, or creode, exhibits a balance between inflexibility (tendency to reach the normal end-result in spite of abnormal conditions) and flexibility (tendency to be modified in response to circumstances). I have used the word 'canalization' to refer to this limited responsiveness of a developing system. The course of differentiation tends to follow its normal path, which can be imagined as lying along the bottom of a valley; but it can be forced out of the valley-bottom and some way up one side by abnormal conditions, provided they are strong enough.

The concept of a creode brings together considerations which had previously been placed in two separate bodies of theory, one of which dealt with the effects of genes on development and the other with the effects of the environment in modifying the characters of the organism. Geneticists have, ever since the beginning of their science, recognized that some genetical alterations are ineffective in producing changes in the organism. For instance, if a certain species of animal normally contains two genes AA, one of these may change into an A′ and it may be found that the animal containing A′A looks exactly or nearly the same as one with AA. The gene A′ is then said to be recessive and A dominant. Looking at this from the developmental point of view, one would say that the creode leading to the character affected by A and A′ is so well canalized that the change of one A into an A′ is not sufficient to divert development to some abnormal end-result. The fact that the creode is an expression of the potentialities derived from the whole set of genes is well illustrated when one finds that if certain other genes are also changed, the alteration of one A into an A′ may become more effective. As the geneticists put it, the degree of dominance is influenced by the whole set of genes; as the developmentalist would say, the degree of canalization of the creode is a function of the whole set of genes. Again some changes in environment may be ineffective in causing alterations. Embryos developing in mildly abnormal temperatures, or water with slightly unusual salt content, may finish up quite normal, while greater variations in these factors may result in abnormal adults. Here again we are dealing with the degree of canalization of the creode. We shall see in the next chapter the importance for evolutionary processes of the fact that changed genetic endowments and changed environmental circumstances both operate by affecting the same basic developmental creodes.

Thus our theory of the development of different types of tissues is atomistic in so far as it involves the presence of atomistic genes, but it is of a continuum type in so far as it regards these genes as carrying out activities which interact with one another in such ways that they are organized into systems. There need be nothing very mysterious about their interaction. Since the days when these ideas about development were first propounded, some twenty years ago, the study of interacting systems has grown up into a well-recognized branch of scientific theory, now usually known as cybernetics. It may be worth mentioning two types of cynbernetic interaction which might underlie the interactions of gene activities to form systems of the kind with which we are concerned (Fig. 7).

Suppose that in the cell there are two possible sequences of chemical change; from P to Q to R and then on to something else, or from A to B to C and so on. Now suppose that the presence of Q prevents the conversion of A to B, while the

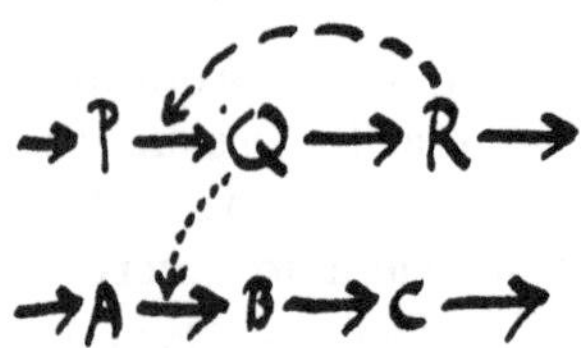

FIG. 7

Two alternative chains of reactions, with feed-back (dashed line) and interaction (dotted line).

presence of B prevents the conversion of P to Q; then these two paths will be alternative. A cell can either be doing P to Q to R or it can be doing A to B to C, but it cannot be doing both simultaneously. Again, suppose that the rate at which Q is formed is reduced by the presence of R. This will act as a sort of governor, so that if the cell does P to Q to R at all, it can only do it at a definite rate, since as soon as the reaction starts to go too fast, too much R is produced and this slows it up again.

I am mentioning these excessively simple models only to show that the interaction of gene activities to form organized systems does not involve any very surprising or unknown properties, but is something which we have a perfect right to expect.

Actually we have still very little definite knowledge of exactly how these regulatory systems operate in practice. For instance, when we speak of some gene being, in certain tissues, more fully 'activated' than others, is the activation to be thought of as something positive or negative? Is it due to an addition of something without which the genes cannot work, or is something removed which was previously inhibiting the gene from being effective? Perhaps the latter is the simpler alternative, and there is some evidence that it is the lifting of an inhibition, rather than a positive activation, which occurs in such phenomena as 'enzyme adaptation'; but we still have no clear answer to the question as far as normal embryonic cells are concerned. Again, exactly how are a set of gene activities tied together into an organized group? It might be that this is due simply to their dynamics—to such factors as competition for substrates, mutual facilitation and inhibition, and so on. Biochemistry has not yet developed techniques adequate for studying this matter.

It is still at the stage of sketching in the broad outlines of the processes of synthesis; for instance, in discovering the mechanisms by which the necessary energy is made available. It is already known that RNA is involved in the synthetic processes in some way, and presumably there must be different specific RNA's corresponding to the various proteins, but as yet biochemistry is not in a position to identify them individually. It can therefore not fill in the detail of the picture whose general outlines emerge from the genetical considerations.

Another line of evidence emerges from the microscopic investigation of the structure of cells as they are engaged in processes of differentiation. The electron microscope has in the last few years revealed to us a whole range of structures, which lie in the scale of magnitudes just greater than those dealt with by the chemist, but less than those seen with the light microscope. Such studies are in their infancy, but it appears already to be emerging that a cell, at the time it starts to manufacture the characteristic proteins which will form it into a nerve cell, muscle cell and so on, undergoes alterations of internal structures of this range of magnitude.* These affect particularly the envelope which encloses the nucleus, and the systems of flattened vesicles or double membrances which compose the so-called endoplasmic reticulum in the cytoplasm. Moreover, these structural changes seem to be characteristic of the particular direction in which the cells are differentiating. It seems quite likely that much of the interaction between the synthetic processes, by which they become coupled together into organized systems, takes place through the formation of these relatively large structured entities; and in that case it could not be fully described in purely chemical terms. This remains, however, for the future to determine.

This discussion so far has only dealt with the development of parts of the egg into cells of definite types, and we have left on one side the phenomena of individuation by which these cells become arranged into organs with definite shapes and patterns. I am afraid biologists have to confess that they still have hardly any notion of how this is done. It certainly must involve something more than purely chemical processes. Development starts from a more or less spherical egg, and from this there develops an animal which is anything but spherical; it has arms, legs, head, tail and so on, and internal organs which also have definite shapes. One cannot account for this by any theory which confines itself to chemical statements, such as that genes control the synthesis of particular proteins. Somehow or other we must find how to bring into the story the physical forces which are necessary to push the material about into the appropriate places and mould it into the correct shapes.

We can still only make very vague suggestions as to how this might be done. For instance, one possible type of process is as follows. One of the simplest examples of the individuation of a mass of cells into an organ of a definite shape is the formation of the notochord.* This is a long rod-like organ which is the first rudiment of the future back-bone. In the newt embryo it is formed out of a large number of cells, which are at first roughly spheroidal in shape and arranged in a haphazard manner as a rather broad sheet. These cells gradually draw themselves together into a more compact, elongated mass in which the cells have roughly the shape shown in Fig. 8. This then gradually

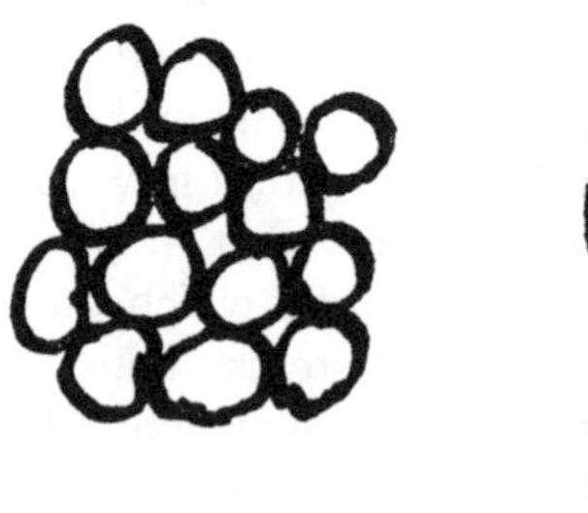

FIG. 8

The shape of the cells in the notochord of a newt, from an early stage on the left to a later one on the right.

changes into a more elongated rod in which the cells have become flattened, and are lying up against one another, as it has been said, rather like a long pile of pennies. Finally each cell becomes blown out by the appearance of a large vacuole inside it, and this leads to a still greater elongation of the whole structure. Now this series of changes in the shapes of the cells and of the structure which they build up, could conceivably be explained if we supposed that the cells changed in such a way that the region in which two cells were in contact with one another became enlarged. At first a cell touches rather loosely a lot of neighbours. If the largest area in which it was in contact with a neighbour increased in size, this would mean that it would have to lose contact with certain other neighbours which it had previously just touched, and eventually the mass would be drawn into the form of the pile of coins in which each cell has the maximum possible contact with two others, one on each side of it. Even the swelling of the cells by the appearance of the vacuoles inside them could be regarded as a way of increasing the area of contact between neighbours. Here the electron microscope comes to our help, and has allowed us to see that there really is such a tightening-up of the cell-to-cell contacts, although, of course, it does not make it possible to decide whether this is the cause, or only a symptom, of the changes in shape which we see. But, whether the cell contacts are the most important operative factors or not, it is clear that the development of shape must be explained by some theory involving physical forces and not merely chemical activities (although, of course, the physical forces will be produced between systems which have certain chemical characteristics).

Whatever the exact nature of the physical forces which mould the developing tissues into organs, they must certainly be organized into self-regulating systems. It is one of the most striking characteristics of embryos that they 'regulate'; that is to say, if pieces are cut out of them or they are injured in various ways, they have a great tendency nevertheless to finish up by producing a normal end result. The forces by which the shapes are produced cannot be working quite independently and without regard for what is going on around them. They must be showing the same capacities for compensating for abnormalities and accommodating themselves to their surroundings as we saw in the case of the chemical processes involved in the differentiation of tissues.

Thus, when we consider development, we find that our theories have to involve several levels of organization over and above the simple atomicity of thinking in terms of sequences of DNA and of protein. We have to deal first with cybernetically organized systems of gene activities, which define a set of alternative creodes, along which various parts of the egg change until they develop into well-defined types of cells. Then we need to postulate organized systems of physical processes which bring about the individuation of these groups of cells into organs which have definite shapes and sizes. There may be other still higher levels of organization. For instance, there are remarkable phenomena of 'regionality'. In the embryo chick, at quite an early stage, it can be shown that the hind limbs differ in some definite way from the fore-limbs, although both are at this time only small lumps of apparently identical-looking cells. However, if a small fragment is cut out from the upper region of the leg and placed near the tip end of the wing, it has been found in some cases to develop in this new situation into a claw. This means that at the time the grafting was done it was still undecided which region of the limb the cells would develop into. They were taken out of the upper region (i.e. the thigh) and have become the distal end (i.e. a toe); but still they have retained the quality 'hind-limb' as opposed to 'fore-limb', so that, although they are now developing in a wing, it is not the distal end of a wing that they become, but the distal end of a leg. We have still not the slightest idea how to interpret the nature of such a regional quality. Is there some chemical difference between all parts of a leg, whether thigh or claw, and all parts of a wing? We simply do not know, and it is possible that here we are seeing a still more complex level of organization than those which produce tissue differentiation and organ individuation.

It is, of course, only a beginning of understanding to say that the processes we are investigating force us to think in terms of theories which involve organization. Where does this organization come from? How does it come about that the gene activities are such that they interact in a way which gives rise to sharply alternative creodes? It is too much to suppose that if we took just any arbitrary set of genes, each capable of controlling the synthesis of a particular protein, and put them all together in a single cell they would necessarily interact with one another in this way. But, of course, in biology we are not dealing with arbitrary groups of genes which have been assembled purely by chance. The organisms we see in the world around us have been produced by the processes of evolution, their organization—and that means the organization of their gene activities so as to define creodes and of their morphogenetic forces so as to produce organs with a definite pattern—must have been brought into being by evolution. In the next chapter, therefore, we shall have to pass on to consider the still longer term set of changes which living organisms undergo.

CHAPTER 4

*Evolution**

WE came, in the course of the last chapter, to look on living things as dynamic developing systems which are organized by means of feed-back or cybernetic relations. This organization we believe to have been produced by a process of evolution. The scientific theory of evolution dates, in most people's estimation, from the time of Charles Darwin. This makes it almost exactly 100 years old; since it was in 1859 that his book, *The Origin of Species by means of Natural Selection*, was published. In the century of its existence the theory has had an effect perhaps more profound than that of any other scientific theory whatever on man's general way of thinking about the universe in which he lives and about his own nature. As would be expected, any set of ideas which are so enormously influential cannot be very simple, but need considering from a number of different points of view; and in any discussion of evolution there are bound to be many strands of meaning which have to be kept in mind. It may be as well to begin by mentioning some of the main foci of interest around which the rest of this chapter will revolve.

The simplest point at issue in connection with evolution is whether it occurred. This is the question whether the species of animals and plants which we see living in the world today originally came into being in their present form, or whether they have only reached their present condition through a process of transformation from other types which existed at earlier periods. Has the living world really undergone a process of progressive change—a change which Darwin usually referred to as 'development' or 'transformation', but which we nowadays speak of as 'evolution'? The usual, and perhaps the only, alternative view is to suppose that each species was brought into being by a special act of creation at some particular point in history.

A second question is whether the living beings of the world have been brought into existence by the activity of something which can be validly compared to the creative intelligence, or alternatively whether they have assumed their present form by the operation of 'natural' processes not involving anything comparable to intelligence. Darwin's Victorian contemporaries almost without exception regarded this question as essentially connected with that asked in the last paragraph. They seemed to have assumed that if an intelligence had operated it must have done so through a series of separate acts of creation. We shall see later that this assumption has not always been made.

This brings us to the third major group of questions which may be asked about evolution. Supposing that some type of evolution has occurred, and that transformation of species has really taken place, what type of causal processes must we suppose to have been active? The notion of causation which has, I suppose, been conventional in most human thinking until fairly recently, envisages a cause, or perhaps two or three causal agents a, b, c which produce some effect X. Darwin's theory was one of the first important scientific achievements in modern times to place its main reliance, not on simple causation of this type, but on processes of the kind we usually refer to as chance. The success of his theory has had a profound effect in making probability theories, or stochastic thinking, respectable. Much of modern physics is now phrased in such terms, and so at the other end of the spectrum of sciences is most of sociology. But I shall want to argue that in the sphere of evolution we are now finding ourselves confronted with the need for organismic thinking. Darwin's emphasis on the importance of chance was a crucial step in breaking the hold over men's minds of the notion of strict causal determinism of the Newtonian kind; but once this idea has been abandoned, the idea of organization, or cybernetic schemes of causation should be recognized as just as important as that of random chance, or stochastic causation.

With this rough map of the country to be surveyed in our minds, we can now begin to look at the theory of evolution in somewhat greater detail. To see it in proper perspective we should begin long before Darwin's time. Many of the major problems of the natural philosophy of evolution had been formulated ages before he wrote. The astonishing Greeks, who seem to have opened and poked their head out of almost every window of the mansion provided for man's habitation on earth, cast their eye over this landscape also. The great atomic theorists, Empedocles, Democritus and their later followers such as Lucretius, envisaged a world in which evolutionary changes had taken place as a result of the chance collisions and comings together of atoms. In opposition, Anaxagoras saw such changes as resulting from intelligent design.

Darwin, in the 'Historical Sketch' which he added to the later editions of his *Origin of Species*, argued that the most famous of all the ancient biologists, Aristotle, adopted the atomistic or Lucretian view, which was the nearer to Darwin's own. However, as Basil Willey has recently pointed out,* this was an error on Darwin's part. Aristotle repeated the atomistic arguments only to repudiate them. His conception of nature was one which we no longer entertain in any serious form; it was the 'teleological' view. This supposes that existing things come into being in relation to a formulated design, but the design is placed, as it were, after them rather than before. It is not thought of as a will which attempts to create them, but as an end which they strive to attain. This point of view led Aristotle to classify living organisms into a hierarchial system, but one which he thought of as a static arrangement rather than as representing a progression through which the living world had moved with the passage of time. Plants were 'lower' than animals; sponges, jelly fish and so on 'lower' than worms; these again were surpassed in perfection by the other types of animals, until one reached the 'highest' groups and so to man himself. But to Aristotle this arrangement was a classificatory system in terms of degrees of perfection and not in terms of historical succession.

Throughout most of European history, from the time of the Greeks until the end of the eighteenth century, men's ideas about the world of living things were a compound of Aristotle's repudiation of atomism, his recognition of a hierarchical system of living things, and, usually, a literal acceptance of the Book of Genesis as a statement that the living things we know had been created at a single point in time. A beautiful expression of these thoughts is given by Milton in *Paradise Lost*—incidentally, Milton was the one author whose works Darwin always took with him on excursions during his voyage on the *Beagle*.* Milton's description of the system of living things comes in Book 5. Raphael has been sent by God to visit Adam and Eve, who were then still in the Garden of Eden. The scene opens with Eve, like a good and thoughtful housewife, preparing lunch for her husband and his guest:

> So saying, with dispatchful looks in haste
> She turns, on hospitable thoughts intent
> What choice to choose for delicacy best,
> What order so contrived as not to mix
> Tastes, not well joined, inelegant, but bring
> Taste after taste upheld with kindliest change:
>
> Raphael eats with keen dispatch
> Of real hunger, and concoctive heat
> To transubstantiate . . .
>
> Meanwhile at table Eve
> Ministered naked, and their flowing cups
> With pleasant liquors crowned. O innocence
> Deserving Paradise! If ever, then,
> Then had the Sons of God excuse to have been
> Enamoured at that sight. But in those hearts
> Love unlibidinous reigned, nor jealousy
> Was understood, the injured lover's hell.

Adam then asks about the nature of the world into which he has come and Raphael explains:

> O Adam, one Almighty is, from whom
> All things proceed, and up to him return,
> If not depraved from good, created all
> Such to perfection: one first matter all
> Endued with various forms, various degrees

Of substance, and, in things, that live, of life;
But more refined, more spiritous and pure,
As nearer to him placed or nearer tending
Each in their several active spheres assigned,
Till body up to spirit work, in bounds
Proportioned to each kind. So from the root
Springs lighter the green stalk, from thence the leaves
More aery, last the bright consummate flower
Spirits odorous breathes; flowers and their fruit
Man's nourishment, by gradual scale sublimed,
To vital spirits aspire, to animal,
To intellectual; both give life and sense,
Fancy and understanding; whence the Soul
Reason receives, and Reason is her being, . . .

In this passage, with its wonderful metaphor of the growing plant—I like to think that Milton had in his mind's eye the garden of Christ's College at Cambridge, which cannot have altered so much in the quarter-millennium between the time Milton was a Fellow there and the years when I had rooms looking out on it—there is some suggestion that 'flowers and fruit' have gradually become changed into animal and intellectual spirits, by a progression which one might think of as evolutionary. But this cannot have been what Milton intended to imply. In the seventh Book of *Paradise Lost* he gives a clear picture of Creation as a single act:

when God Said,
Let the Earth bring forth soul living in her kind,
Cattle, and creeping things, and beast of the earth,
Each of their kind! The Earth obey'd, and straight
Opening her fertile womb teem'd at a birth
Innumerous living creatures, perfect forms,
Limb'd and full grown . . .

. . . now half appear'd
The tawny lion, pawing to get free
His hinder parts—then springs, as broke from bonds,
And rampant shakes his brinded mane; . . .

However, it is important to realize that the idea of the evolutionary transformation of living things was not completely repudiated even in the Middle Ages. As Basil Willey has recently reminded us, both St Thomas Aquinas and St Augustine held that 'in those first days God made creatures primarily or *causaliter*'; and by this they meant that he created not the finished creatures but potencies such that the earth would gradually give rise to them by a process of unfolding. Accordingly, Aquinas continues, after the primary Creation God 'then rested from His work, and yet after that, by His superintendance of things created, He works even to this day in the work of propagation'.

The possibility of such a point of view seems to have been almost entirely overlooked in Darwin's time. Most of his contemporaries, both within and without the Churches, found it very difficult to reconcile any form of transformationism with a belief in the reality of a Creation by God. The argument that evolution had occurred was taken to lead to an atheistic conclusion. It took some decades at least for the organized Churches to realize that this is not necessarily so. Acceptance of evolution does make it impossible to believe in a God who operates in the way which a literal interpretation of Genesis suggests; but I think most biologists would agree with the modern theologians who argue that interpretations do not have to be literal, and that the question of Theism versus Atheism cannot be settled by reference to facts about the entities we observe in the world around us.

There was another aspect of Darwin's theory which, to his contemporaries, seemed even more irreconcilable with belief in an intelligent Creator. That was his atomism, and his reliance on chance as an essential part of the evolutionary process. Newton's *Opticks*, published at the very beginning of the eighteenth century, had propounded his enormously influential corpuscular theory of light, and this had raised a renewed interest in atomic theories in general, including the old Epicurean or Lucretian heresy that the basic process of the universe is the chance collision of atoms. Newton himself had enquired: 'How came the Bodies of Animals to be contrived with so much Art, and for what ends were their several Parts? Was the Eye contrived without Skill in Opticks, and the Ear without knowledge of Sounds?' His own answer, of course, held that these organs were evidence of the activities of an intelligent Creator. But there were others, the heirs of Descartes as well as of Newton's own scientific writings, who were attracted by the Lucretian mechanisms. As might be expected, they were in general rebuked by the literary circles of the day,* as when Blackmore wrote:

> Lucretians, next regard the curious Eye,
> Can you no Art, no Prudence there descry?
> By your Mechanic Principles in vain
> That Sense of Sight you labour to explain.

And Matthew Prior in *Alma* also pointed out some of the difficulties that confront attempts to explain the formation of organized structures without bringing in considerations concerning their use:

> Note here, Lucretius dares to each
> (As all our youth may learn from Creech)
> That eyes were made, but could not view,
> Nor hands embrace, nor feet pursue;
> But heedless Nature did produce
> The members first, and then the use.

Here, even before the theory of evolution had been developed to a state in which it was an important influence on thought, a question had emerged which was found to become crucial as soon as transformation of species became accepted as a fact. If animals and plants evolve—whether or not they were initially created *causaliter*—what type of causation operates in the processes by which they alter?

Orthodox thinkers sought for some kind of simple causation—one cause producing one effect. For those content with a religious category of thought, the one cause was the will of God operating throughout the course of the earth's history. For those who looked for a scientific explanation and enquired after the nature of the evolutionary processes occurring within the world as we perceive it, the most important theory put forward during the eighteenth century was that of Lamarck. He supposed that evolution had come about by means of processes which involved causal interactions between the organism and its surrounding circumstances, the environment. The first stage in an evolutionary change is for the organism (Lamarck was thinking of animals rather than plants) to decide by an act of will to change its environment—to move into a new region or to carry on its life in some different way from that it had used in the past. Its new habits created what Lamarck called new 'besoins', and new structures then arose in the animal in correspondence with these. The French word 'besoins' has often been translated as desires or wishes, so that many people understood Lamarck to be claiming that in evolution new structures are brought into being by the desires of the animal. As Graham Cannon* has recently pointed out, a more just interpretation is 'needs'. Lamarck was arguing that when an animal is faced with new necessities in the carrying on of its life it will develop new structures or abilities suitable for performing what is being required of it. Moreover, Lamarck urged that these new structures or facilities would be passed on to the offspring through heredity so that they would result in a true evolutionary change.

The processes envisaged by Lamarck depend on a normal and well-recognized type of causality, but from the point of view of his contemporaries his theory suffered from one great defect, and from that of his successors from another. The first thing that was held against him was that his ultimate cause is to be found in an act of will. This, in his scheme, was attributed to the animals concerned, but in its nature it was something of the same kind as the creative will which the theists attributed to the Deity. Those who were ready to accept the transformation of species were repelled by the invocation of a psychic, if not a spiritual, basis for the process; while those who might have found such a basis acceptable, either denied that evolution had occurred, or felt that if it had, no more was necessary to explain it than the will of God. The second argument against Lamarck was, of course, the thesis, generally accepted at the present day, that characters acquired by an organism during its life-time, whether as the result of an act of will or in any other way, are *not* in fact passed on to the offspring, and therefore cannot take a part in the evolutionary process.

This second argument, however, did not acquire much force until well after the time of Darwin's work. Darwin, in fact, by no means rejected the inheritance of acquired characters, or, as he put it, of the effects of use and disuse. His repudiation of Lamarck was on the first account. He could not stomach such vitalistic notions as the effectiveness of an act of will and the calling into being of new organs by the needs of the organism.

Darwin's great achievement was to put forward a mechanism by which evolution might be brought about without requiring the operation of such doubtful agencies. In doing so he returned wholeheartedly to the Epicurean and Lucretian reliance on chance. He pointed out that individuals of the same species differ from one another by random variations which are to some extent hereditary. Then he goes on to argue that all organisms produce many more offspring than can possibly survive to make up the next adult generation. Out of all those which are born, the actual survivors will be selected by the natural chances and hazards that they encounter as they mature. On the average, however, those parents which vary from the norm of their species in certain ways—for instance, by being stronger, more resistant to disease, better at holding their own against predators, or escaping from them and so on—will leave more offspring than those which vary in the other direction. Thus the frequency of these favourable variations will be slightly higher in the next generation, and will in fact continually increase as the generations pass. It is, according to Darwin, this increase in the frequency of favourable variations that constitutes evolution.

Writing a hundred years ago, Darwin had to leave a major gap in his theory, at the point where he should have explained how the random variations arise in the first place. There was at that time no theory of heredity, no understanding of the nature of hereditary factors, and therefore no possibility of arriving at any clear understanding of how they may change. With the development of genetics, which we glanced at in the second chapter, this lacuna in the theory has been filled. We now know that heredity is carried from one generation to the next in the form of separate hereditary genes, each of which can be looked upon as an exceptionally complicated molecule of nucleo-protein. Few biologists, even today, would claim that we have any full understanding of how such genes may change, or 'mutate'. We know, however, that they do so spontaneously, and we have many reasons for believing that a large proportion of these spontaneous changes, if not all, arise simply from the fact that such very large and complicated structures are inherently somewhat unstable. If any moderately large collection of atoms becomes arranged, by the formation of bonds between the atoms, into a molecular grouping, there are likely to be a number of alternative different types of bonds, and therefore of molecular groupings, which are possible. It is by no means unexpected to our present outlook to suppose that the exceedingly complex groupings which constitute genes may sometimes alter from one of their possible forms into another for no definitely ascertainable reason. There is indeed more reason to be surprised at their stability than at their tendency to mutate. One might almost have expected structures as complex as this to show some of the properties of a house of cards, which may collapse entirely with nothing more than a slight breath of wind or a jog at the table. Most biologists would agree, therefore, that when Darwin spoke of the variation between individuals as random, he made at least a first approximation to the truth.

Our present theory of evolution can indeed be regarded as for the most part no more than a restatement of Darwinism in terms of Mendelian genetics. I shall not have time to give more than a very cursory summary of it here. It falls, one may say, into three major parts. The first is the problem of the origin of new hereditary variation, which we have just discussed. The second is the elucidation of what is involved in natural selection. The third is the analysis of the evolutionary situation existing in populations of animals and plants as we meet them in nature.

The history of the idea of natural selection is a somewhat peculiar one. The first geneticists in the early years of this century were little interested in the notion. They were exceedingly impressed by the newly-discovered facts about the inheritance, according to Mendelian rules, of new hereditary characters and, naturally enough, they selected for study characters which were easily recognizable as departing considerably from the norm of the species. This encouraged them in the impression that the major factor in the evolution of new species is the origin of a new striking hereditary form by means of a gene mutation. To the mechanism of the preservation or spreading of the new form they paid much less attention. They were, in fact, rather antagonistic to Darwin's notion that the basic evolutionary process is the production of *slightly* more offspring by an organism which is *slightly* fitter in some ways than its contemporaries.

It was, more than anyone else, the mathematical theorists who began writing about evolution in the 1920s—Haldane, Fisher and Sewall Wright—who rescued natural selection from this neglect. But in doing so they reduced it to a truism or tautology. Darwin had argued that it is the stronger or the more resistant, or, in general, the 'fitter', organism which leaves most offspring. But if one tries to generalise a concept of fitness which will make this statement true, one finds that the only way in which fitness can be defined is by the statement that it is to be measured in terms of the number of offspring contributed to the next generation. The slogan 'the fittest survives', which Huxley had coined to enshrine the kernel of Darwin's theory, and which had seemed to the social Darwinists to have such profound implications for human affairs, had to be replaced by the innocuous truism that those individuals that leave most offspring will make the greatest hereditary contribution to the next generation. A thesis is, of course, none the less true and important because, once it is stated, it is obvious. But when we see that somethng is a straightforward piece of common sense—as Darwin's theory of natural selection is—we are perhaps less inclined than before to think that it lets us in for the first time to the secrets of the universe, and can be applied as a magic formula to all and sundry problems.

The other major route which has led to our deepening understanding of evolutionary processes has been the study of populations of animals and plants in nature. What kind of variations in hereditary properties can we discover in them? The strikingly unusual hereditary variants, which so impressed the early geneticists, turn out on examination to be usually of little importance from the point of view of evolution. The majority of them are in fact highly abnormal individuals, which are extremely unfit (i.e. leave very few offspring) in normal circumstances, and which it is difficult to envisage ever being highly effective in any ordinary circumstances at all. However, from the very earliest days of genetics it has been recognized that genes affect even the slight and almost imperceptible variations by which individuals, all of which we should consider normal, differ in minor ways from one another. Very soon after the rediscovery of Mendel's work, the statistician Udny Yule* pointed out that if a character, such as the stature of man, is influenced by many different hereditary factors, then in a population which contains many variants of these genes, individuals will vary by imperceptible degrees over a wide range of heights. There will be a large number with heights around the average, somewhat smaller numbers who are slightly taller and slightly shorter, smaller numbers again who are still taller or shorter, and so on with diminishing contingents as we approach the extremely tall or the extremely short.

Genetical experiments conducted on natural populations show that they do indeed usually contain many variants of sets of genes which affect characters like stature, resistance to disease, and in fact most of the properties which one would expect to influence the number of offspring left by a given individual. We now think that it is with such genes that the natural selection which brings about important evolutionary consequences is mainly concerned. It is only rarely that we can identify a single particular gene with a fairly strong effect as being of importance in evolution. But such circumstances do occur. For instance, the recent blackening of tree-trunks and vegetation by industrial smoke in certain parts of Great Britain and Europe has led to the natural selection of dark coloured forms of moths, which can easily conceal themselves against these murky backgrounds. Similarly the application of insecticides on a mass scale has led to the appearance of resistant strains of insects such as mosquitoes, and flies. In both these cases, a large part of the effect has been produced by one or two genes with comparatively strong and easily recognizable effects. Usually, however, the environment does not change so abruptly, and the fittest animals are nearer to the normal of the species; and natural selection is then concerned with characters which are influenced by many genes, each with only a small effect.

In the sixty years since Yule wrote we have, of course, learnt much more about the systems of many genes which underlie continuous variation. In a recent authoritative summary of modern genetical theory,* Ernst Mayr suggests that we have seen two consecutive outlooks on the genetics of populations. The first concentrated its attention on the behaviour of single genes of strongly marked activity; while in the second 'not only individuals but even populations were no longer described atomistically, as aggregates of independent genes in various frequencies, but as integrated, co-adapted complexes'. Mayr has referred to this new mode of thinking as the genetic 'theory of relativity'. I doubt myself whether these two periods can be distinguished as clearly as Mayr suggests. Admittedly, in the earliest days some of the most important geneticists discussed evolution in terms of single identifiable genes; but there were always others, not so fashionable at the time, who were less atomistic in their thinking. From Udny Yule through Weinstein, Fisher and Sewall Wright and their many modern followers the interest in systems influenced by many genes has been continuous from the earliest times.

The phrase I have quoted from Mayr puts its finger on one of the most important developments which our knowledge in this connection has undergone in recent times. Students of natural populations, of whom perhaps the most important is Dobzhansky, have shown that in order to account for the effects of genes on biological fitness we have to consider them, not each for its own sake, but as forming co-ordinated groups. Suppose that in a given population there are the genes A a, B b, C c, D d, and so on, each gene existing in two forms as indicated by the large and small letters. We may find that certain particular combinations, for instance, A A, B b, c c, D D and A A, b b, C c,

D D are particularly fit, while all the other combinations give organisms which in competition with the former types, leave many fewer offspring. In some slightly different circumstances the particularly fit combinations may be somewhat different—perhaps a a, B b, c c, D D and a a, B b, c c and D d. Under these circumstances we will find that in populations in the first situation there will be a large number of the genes A, B, b, C, c and D, but hardly any of a, or d, while populations living in the other situation will have many a, B, b, c, D and d, but very few A or C. Any population of these organisms will contain a particular collection of genes, or 'gene pool', which is limited to those which can combine with each other to give tolerably fit individuals. It will include enough different varieties of genes so that a large number of different combinations can be produced, resulting in hereditary variation between the individuals, but it will exclude, or keep at extremely low frequencies, those genes which fail to fit in with the rest to form efficient individuals. This is what is meant by saying that the genes in the gene pool of any given population are 'co-adapted' to one another. The discovery of the phenomenon of co-adaptation and its study is the furthest point to which the genetical analysis of populations has yet reached.

These developments which have arisen from the application of modern genetical theory to the study of evolutionary processes do not alter essentially the basic feature of Darwinist thought, which is its reliance on chance rather than on a simple determinist type of causation. Thus, in a later part of the lecture already referred to, Mayr writes 'What do we mean by twentieth century Darwinism and what do we mean by the synthetic theory of evolution? I think its essence can be characterized by two postulates: (1) that all the events that lead to the production of new genotypes, such as mutation, recombination and fertilization, are essentially random and not in any way whatsoever finalistic, and (2) that the order in the organic world, manifested in the numerous adaptations of organisms to the physical and biotic environment, is due to the ordering effect of natural selection'. This emphasis on the importance of chance has been one of the most profound and far-reaching of Darwin's influences on human thought. It spread into fields far removed from those which Darwin discussed. As we all know, during this century there has been a strong tendency to frame the laws of physics in terms of probability or chance events, rather than in terms of the type of simple causation which had been relied on by Newton.

Within the field of evolution the rival type of hypothesis, which the reliance on chance superseded, was one which depended on the operations of an intelligent designer. Darwin himself, to some extent at least, shared the feelings of many of his contemporaries, that the substitution of chance for design as an explanatory principle tended to undermine one of the major intellectual reasons for a belief in God. 'I may say', he wrote in one of his letters,* 'that the impossibility of conceiving that this grand and wonderous universe, with our conscious selves, arose through chance, seems to me the chief argument for the existence of God; but whether this is an argument of real value, I have never been able to decide. . . . The safest conclusion seems to be that the whole subject is beyond the scope of man's intellect; . . .'

Many of his readers, particularly those who were not scientists, could not bring themselves to adopt such a neutral attitude, and felt deeply shocked. As Irvine has put it,* 'Darwin's explanation of evolution is mechanistic without the favourable implications of mechanical design. Natural selection represents not a harmony but a conflict, and is effected not by the precise, mathematical idealism of invisible force, but apparently by a crude, random sorting out of variations by the environment. . . . Many who were willing to believe in an evolving Deity could not believe in one who dealt in random variations. They could accept an evolving universe but not a universe shaken out of a dice box.'

These feelings have not been effective in halting the triumphal progress of theories relying on chance, either in biology or in the other sciences. Mayr is quite justified in stating that present day evolutionary theory is firmly based on the postulate that the processes of mutation, recombination and fertilization are essentially random. Moreover, voices have been raised against the view that this renders them uninspiring and barren of spiritual content. The statistician R. A. Fisher, for instance, argues that simple causation of a strictly deterministic kind cannot have results which can be regarded as creative.* In a strictly determinist universe uninfluenced by chance it would be possible, from a knowledge of the present situation, to calculate exactly what would occur at any later time. There would be no room for anything which could be regarded as essentially new and thus as a result of a creative process. It is, he argues, only because the causal processes in the actual world have an essentially chance element in them that we can find any situations to which the concept of creation can properly be applied. 'Natural causation', he writes, 'has a creative aspect . . . because it has a casual aspect . . . Looking back at a cause we can recognize it as creative; it has brought about something which could not have been predicted—something which cannot be referred back to antecedant events. Looking forward to it as a future event, there is in it something which we can recognize as casual. It is viewed thus like the result of a game of chance; we can imagine ourselves able to foresee all its forms, and to state in advance the probability that each will occur. We can no longer imagine ourselves capable of foreseeing just which of them will occur.' Now it is probably true that nothing worthy of being considered creative can occur in a fully deterministic universe ruled by the operations of simple causation. It was a major service of Darwinism—though one that he probably did not anticipate and possibly would not completely approve—to have broken the hold on our minds of notions of simple causation. On the other hand it does not seem to me by any means clear that the concept of creation is necessarily appropriate to any and every other system of thought which does not rely on simple deterministic causation. Admittedly the idea of creation would be out of place in a world which merely ran on like a wound-up watch, but is there really any more room for it in a world which was nothing but a series of throws of dice?

But the question is, to my mind, whether we have not tended to over-emphasize the importance of chance processes in evolution, and perhaps in other domains also. Let us begin to consider this by returning to Mayr's phrase quoted above. He wrote that 'all the events that lead to the production of new genotypes, such as mutation, recombination, and fertilization are essentially random and not in any way whatsoever finalistic'. What does this 'finalistic' mean? It is a word that has 2,000 years of theory behind it, going back to Aristotle and his conception of a final cause—a design towards which events in this world are tending. We can now, I think, see an inadequacy in the notion, which has only fairly recently become evident. We have learnt how to construct mechanisms, operated by means of processes none of which are brought about by final causes, but which combine to give a piece of equipment which results in the attainment of some pre-designed end.

Consider for instance an automatic target-tracking gun sight, such as were commonly used against aircraft in the last war. This is a machine which sends out a signal, which becomes reflected back to it from the target; and this is then received by another part of the mechanism which causes the gun to point in such a direction that if it is fired it will hit the target. If such a machine is operating correctly it is quite immaterial when or how the trigger is pulled. This may be done by some purely random event, but whenever it is done the target will be struck. Although none of the elements in the mechanism are finalistic, and one of the most essential features in its operation, namely the time at which it is fired, is entirely under the rule of chance, yet, the equipment as a whole is finalistic in effect. One might call such mechanisms 'quasi-finalistic'. The possibility of their existence means that it is no longer adequate merely to state that a system depends on the operation of non-finalistic processes, or that it is influenced by chance. Even if this is granted, we have still to ask the question whether the total mechanism is such that it has a quasi-finalistic character.

The time has come to ask this question about evolution. Mayr felt that simply proceeding from the consideration of single genes to that of systems of genes, was a step important enough to be dignified by the name of the 'genetic theory of relativity'. We need now, however, to go to a much more far-reaching relativistic theory which brings into the system not merely organized groups of genes but the environment as well.

As soon as we try to do this we find ourselves passing out of the sphere of atomistic theories, whether they deal in simple deterministic causation or probabilistic causation, into the domain of organismic or cybernetic thinking. The first point that confronts us, for instance, is that before an organism's environment can exert natural selection on it, the organism must select the environment to live in. If we release a rabbit and a hare—animals which look rather like one another—in the middle of any ordinary piece of country, the rabbit will run to a hedge or bank and take refuge in it, while the hare will set up house somewhere in the middle of an open field. Even within a single species different individuals differ hereditarily in their behaviour; for instance, in their choice, out of a number of alternatives of an environment to live in, or a member of the opposite sex to mate with. Thus the animal's hereditary constitution influences the type of natural selective pressure to which it will be subjected. And then, of course, the natural selection influences the type of heredity which is passed on to the next generation. We are dealing with a feed-back or cybernetic system in which there is nothing that is simply cause or simply effect.

It is difficult to quote actual experiments which demonstrate how the effect of the organism in determining the nature of the environment may influence its evolution—the application of cybernetic ideas to evolutionary theory is quite recent, and little work has yet been done on under its influence. I shall have to illustrate the situation by rather hypothetical examples. We know that many, indeed probably most animals have a 'cryptic' colouration; that is to say, they are clothed in a pattern of colour and tone which makes it difficult for their enemies to detect them. Some animals, by contrast, have 'warning' colouration, which is extremely easy to spot; these are animals whose flesh is distasteful, or which are armed with unpleasant stings, or other effective defences. It is clear that the cryptic colouration is useful as one method of eluding enemies, and it seems plausible to suppose that the warning colouration is useful because it makes it easy for a predator to recognize that here is an animal which it would be more prudent to leave alone.

The point I want to make is that both these types of usefulness only eventuate when the animal concerned behaves in the appropriate way. The cryptically coloured animal must keep to the surroundings in which its camouflage is appropriate; the animal with warning colour must flaunt itself. If a mutation producing warning colouration turned up in a cryptically coloured species, it could be of no use unless the animal changed over to an appropriate form of behaviour. Again, when for some reason the general background changes, a camouflaged species may have to evolve a new cryptic colouration; but this will be little use to it unless it is used 'sensibly'. We have seen an example of this in the evolutionary changes of moths in industrial areas. The tree trunks on which the moths spend much of their time were, in pre-industrial areas, light in colour; so were the moths, and they were therefore difficult to see. More recently, soot from industrial smoke makes large patches of darkness on the trees; and a dark form of the moths has become favoured by natural selection, because it allows the moths to be better concealed *when they sit on the dark patches*, though they are extremely visible if they rest on the light patches which still remain. It has been shown that the dark moths do indeed preferentially rest on the blackened parts of the trees.* The effective environment in which they are subjected to natural selection is, in fact, the darkened bark which they themselves choose; it is not something completely external, but is a combination of the outside world and the moth's own behaviour.

It is easy to imagine circumstances in which the behavioural component of the environment might be of predominating importance. For instance, the giant Irish elk evolved enormous horns, which were probably used in the fighting between males for the capture of mates. These horns became so large that it is difficult not to conclude that they must have been a great handicap in many of the other activities of life; and this species of elks in fact died out without leaving descendants. It seems likely that the exaggeration in the size of the horns was brought about by natural selection in the quite peculiar environment created by the inflexible insistence of the animals in continuing to indulge in their fighting behaviour. If they had behaved in a different way, other points would have become of greater natural-selective importance, and the handicap of over-sized horns might never have evolved.

We have considerable grounds for believing, then, that mentality in the broad sense, or at least behaviour (biologists tend to be very timid about mentioning the mind), is a factor of importance in evolution. Lamarck's insistence on the 'Will' is not wholly unjustified. But it is not necessary to suppose, as he seems to have thought, that an Act of Will brings into being an appropriate hereditary variation. The situation is that existing modes of behaviour (themselves controlled, with greater or lesser latitude, by heredity) combine with external circumstances to determine the nature of the effective environment.

There is also a second cyclical system which has to be considered in relation to evolution. The environment influences the nature of the adult organisms which grow up within it. When organisms reproduce and leave offspring, the characters which enable some to be more successful than others depend only in part on their hereditary constitution, but in part also on the environmental circumstances under which they develop. The older discussions of the Lamarckian problem of 'the inheritance of acquired characters' usually missed the point that all characters of all organisms are to some extent acquired, in that the environment has played some part—possibly only permissive, but often also to some extent directive—in their formation, and that equally all characters are to some extent inherited, since an organism cannot form any structure for which it does not have the hereditary potentialities. The question we need to ask is not whether acquired characters are inherited, but whether, as we should expect, the ability to acquire the character differs hereditarily in different individuals in a population, and if so what will be the effect of natural selection on the potentialities of later generations. Once the problem has been formulated in this way it is easy to carry out experiments which will give us at least the first answers in this field.

Whenever a population has been tested for the ability of its members to acquire characters during their lifetime under the influence of abnormal environments, it is found that different individuals differ in their hereditary potentialities in this re-

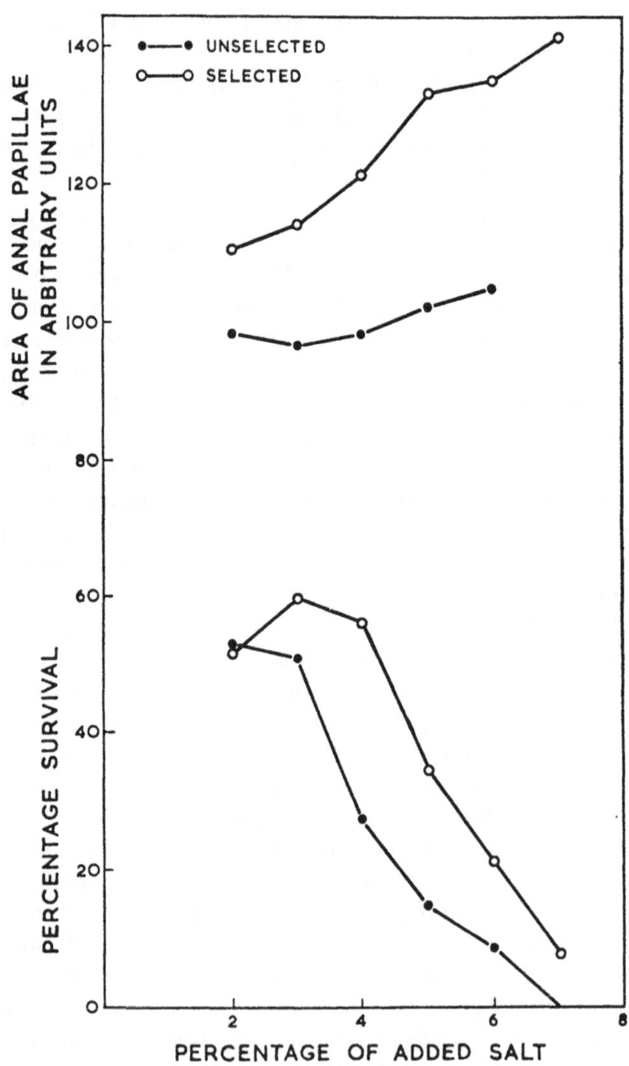

FIG. 9

The effects of growing Drosophila larvae for many generations on food containing added salt. After twenty genera-

spect.* The consequence of this is that if selection, either natural or artificial, operates on the population the hereditary potentialities of the next generation for acquiring the character will be changed. In one experiment, for instance, larvae of the fruit-fly Drosophila were grown on a medium which contained sufficient salt to kill a large number of them. In the unusual environment offered by the medium containing salt, these little fly grubs tend to become slightly modified during their development. The modification takes the form of an enlargement of two papillae on either side of the anus, which are thought to play a part in regulating the salt content of the body fluid. It is, of course, the larvae which succeed in resisting the harmful effects of the salt which survive to give adults which breed and produce the next generation. There is therefore strong natural selection for the ability to survive on the salt. The population was left to the action of this natural selection throughout about twenty generations. At the end of this period a much higher proportion of individuals in it were able to survive on the salt medium.

It was then interesting to discover what had happened to the acquired character, the enlarged anal papillae. In order to study this, a number of larvae were grown on media containing various proportions of salt from 2 per cent up to 7 per cent (Fig. 9) and the size of the anal papillae measured. The results show that the natural selection had brought about two effects. In the first place, the capacity to acquire the character of enlarged papillae had been improved; whereas in the population before the selection operated an increase in salt from 2 per cent to 7 per cent had only resulted in a small increase in the size of the

tions, during which there was strong natural selection for ability to survive under these conditions, the performance of the selected was compared with that of the original strain by growing them both on food containing various concentrations of salt. The lower pair of curves show that at higher percentages of added salt the selected strains survive better than the unselected. The upper curves show the size of the anal pappillae in the two strains; it is larger in the selected strain, and also increases more rapidly as the salt content rises.

papillae, in the population produced by the selection a similar increase in salt content caused a much greater enlargement. In the original population, the ability to acquire this character must have been influenced by a number of different genes. By breeding always from those individuals which were most efficient at acquiring it, all the genes which tended to assist the acquirement of the character would be preserved and passed on to later generations, while genes which were less effective would gradually be eliminated. We finish up, therefore, with a population every individual of which has a hereditary constitution which makes it very efficient at developing large anal papillae if it is subjected to the stress of a high content of salt in its food.

A second result of the natural selection is that the whole level of size of the anal papillae has been raised, not only in the high salt medium, but even on a medium containing a small amount of salt. Thus if we take the population which has resulted from natural selection on the high salt medium and put it back on the low salt food, its anal papillae will still be larger than those of the population with which we started. The acquired character—that is the enlarged anal papillae—has, one might loosely say, become inherited, because it no longer disappears completely when we remove the particular environmental stress which initially brought it into being. We have indeed brought about exactly the same end-result as would be produced by a Lamarckian inheritance of acquired characters. But it is important to remember that this result does not occur because the characters are themselves inherited, as Lamarck suggested. There is no direct transmission of an enlarged anal papillae from an animal which acquired it on a salt medium to its offspring. The apparently Lamarckian end result has been produced through a long course of natural selection operating on the population. To understand exactly how it comes about we need to return to the consideration about the cybernetic systems in development which were considered in chapter 3.

It was pointed out there that developmental processes are organized into systems of such a kind that they show both some tendency to attain their normal end result in spite of abnormal

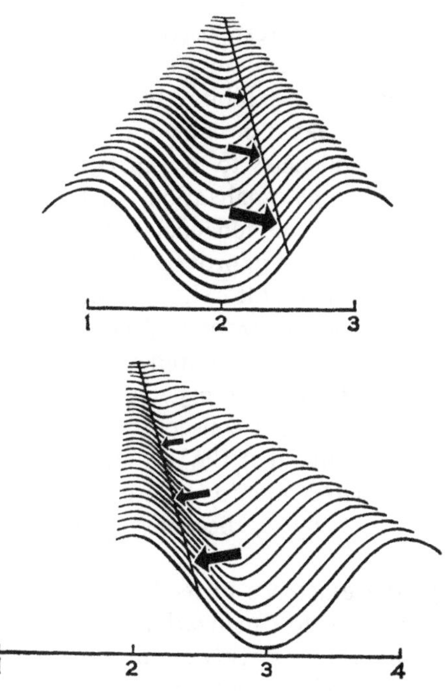

FIG. 10

A diagram of 'genetic assimilation'. The upper drawing symbolizes a developmental pathway which, in normal circumstances, reaches an adult state indicated by the point 2. We suppose that in some unusual environment development can be forced away from its usual course, so that it runs along the line ending at about 2½. If this modification is useful, and favoured by natural selection, we shall eventually select out a population in which development, in this unusual environment, finishes at the state 3, as shown in the lower drawing. If this population is now put back again into the old environment, the relative inflexibility of development may be such that the course of development only goes part of the way back to its original condition, and finishes at about point 2½. Thus some of the 'acquired character' formed in the second environment has been 'genetically assimilated', and now appears as an 'inherited character' in the original environment.

circumstances and some responsiveness to external stresses. The development of any organ—and this includes, of course, the anal papillae—exhibits a balance, between tendencies to flexibility, which make it possible for it to be modified by the environment, and tendencies to go on its own way regardless, which was referred to as canalization. In the salt experiments with Drosophila larvae, the tendencies to flexibility have been, as we have seen, somewhat improved. Nevertheless the development of the anal papillae still exhibits some lack of flexibility, that is, some degree of canalization. When the larvae are grown on medium without salt the development of their anal papillae is not flexible enough to produce organs as small as those of the unselected larvae. It is as a consequence of this restriction in developmental flexibility that the 'acquired character' becomes an apparently 'inherited' one, or to use the phrase I have suggested in this connection, becomes genetically assimilated (Fig. 10).

In this way, the existence of feed-back systems in development, which give rise to canalization, makes possible the appearance of another feed-back system, in relation to natural selection and the environment, which results in the genetic assimilation of acquired characters; and this exactly mimics, by quite another mechanism, the type of result which Lamarck and others have wished to explain by the inheritance of acquired characters.

These two feed-back systems also give rise to still another system of feed-back type, this time in connection with mutation. Consider a population of animals in which there has been natural selection for the ability to acquire some character when they are subjected to a particular environmental stress. In time the hereditary constitution of the individuals will be such that their development is very easily modified by an environmental stress to produce the acquired character. They are set, as it were, on a hair trigger and aimed to hit the target. But once this state has been reached, it will be relatively easy for other things besides the environment to pull the trigger. If genes are changing at random all the time, it will be by no means unlikely that a new mutation will turn up which suffices to pull the trigger, and thus to produce the same acquired character which originally required an environmental stimulus to bring it into being.

To give an actual example: a population of Drosophila was taken and their eggs subjected to treatment with ether vapour. Some of them responded to this environmental stress by having their development modified; they produced very remarkable adults in which the third segment of the body had been modified into a replica of the second segment. Artificial selection was then applied, picking out for further breeding those individuals which responded to the environmental stress in this way. After twenty or more generations, this response also became genetically assimilated; a demonstration that this process can affect even such important characters as the pattern of the major segments of the body, and is not confined only to minor points such as the size of the anal papillae. In our present connection, however, the important point is that during the course of the artificial selection a gene appeared which had a strong tendency, when it was in the population which had been selected, to pull the trigger and release the development of the unusual third segment. If by suitable breeding methods this gene was removed from this population, and introduced into another which had not been selected, it had only a very slight effect in producing the abnormal third segment. It was only when selection had made the developmental system ready to be modified in that manner that the gene was very effective in producing this result. Thus, the selection affected, not, of course, the process of gene mutation itself, but the effects likely to be produced by mutations. The changes which occur in the gene molecules may be random, and yet they may produce results that are not completely random, since the effects of the new genes will depend on the ways in which the developmental system is easily susceptible to modification.

In the evolutionary processes we have, therefore, several examples of organized systems of cybernetic or feed-back relations; one between animal behaviour and environment, and two others affecting acquired characters and mutation, which in their turn depend on the feed-back systems involved in development. The random processes to which Mayr referred operate, then, within a general evolutionary system which exhibits several types of organization. It remains true to say that we know of no way other than random mutation by which new hereditary variation comes into being, nor any process other than natural selection by which the hereditary constitution of a population changes from one generation to the next. But if one confines oneself to the remark that the basic processes of evolution are not finalistic, this, while true, can no longer be regarded as adequate. The non-finalistic mechanisms interact with each other in such a way that they form a mechanism which has some quasi-finalistic properties, akin to those of a target-following gunsight.

There are, of course, many facets of evolution to which these considerations are of little relevance. Evolution has brought about the divergence of, for instance, flies into innumerable species which may differ from one another in ways which appear to us of little interest such as the position of particular bristles on the body, the exact arrangement of the veins on the wings, and so on. However, it is not this type of phenomenon which raises in our minds the problem of design versus chance. It is only certain of the results of evolution which make us feel that something more than pure chance must have been at work, and it is in connection just with these phenomena that we can see ways in which such cybernetic processes as genetic assimilation and the guidance of the effects of mutations might be of importance. Organismic thinking has some contributions to make to evolutionary theory, as a complement to the atomistic outlook, whether that is put in in terms of simple causation or of random chance.

CHAPTER 5

Biology and Man

MAN is, in the first place, a species of animal. As such he is subject, like the rest of the living world, to the processes of evolution which we discussed in the last chapter. It is convenient to analyse the system by which evolution is brought about into four main sub-systems. The first of these, which can be referred to as the exploitive system, deals with the ways in which the organism behaves in relation to its environment and the methods which it employs to keep itself alive. The second, or epigenetic system, is concerned with the developmental reactions by which the fertilized ovum realizes its developmental potentialities by becoming an adult under the particular environmental situation in which it lives. The third is the natural-selective system, which comprises all those processes which determine that some individuals of a species will leave more offspring than others. Finally, there is the genetic system, the series of processes by which information is passed on from one generation to the next in the form of heredity; and this includes the processes of mutation by which hereditary factors may become altered. The effectiveness of these processes did not cease with the appearance of man on earth. He is subject to them as are all other living things.

During the century since Darwin's *Origin of Species* was published, many thinkers have been tempted to seek in their understanding of the methods of evolution for guidance as to how human affairs should be conducted. The early statements of Darwin's theories in terms of 'the survival of the fittest' were used by one school of thought, which was for a time quite influential, to support the idea that human advance must depend on ruthless competition. It was supposed that those who appear most successful in gaining and controlling the goods of this world demonstrated thereby their biological superiority, and not only had, but deserved, a leading role in determining how the future of mankind would develop. With the realization, which was described in the last chapter, that Darwinian 'fitness' is a very special concept, relating to the leaving of offspring and not to any other form of success within the individual's own lifetime, the basis for this theory of 'Social Darwinism' disappeared. It can now be regarded as no more than a temporary aberration in the history of thought.

Somewhat similar attempts to carry over notions derived from our biological understanding into the field of human activity are sometimes made at the present day. For instance, some years ago the school of Soviet biologists who, under the inspiration of Lysenko, believed that heredity may be altered by environmental circumstances rather in the manner suggested by Lamarck, put forward the argument that this had important implications for mankind. They suggested that under a satisfactory social system—which they believed the Soviet system to be—the favourable conditions of life would bring about an actual improvement in the hereditary potentialities of the people fortunate enough to live in that society. They did not, however, draw the equally obvious conclusion that peoples who have lived for many generations under adverse social conditions will have suffered in their hereditary potentialities.

It is perhaps mildly interesting to contrast this view, developed in the Soviet society, with the analogous view that would follow from the application to mankind of the ideas concerning the inheritance of acquired characters that were discussed in the last chapter, which were worked out in a democratic society. According to Lysenko's ideas, an organism subjected to environmental stress merely suffers passively an alteration to its hereditary potentialities—an alteration in a negative sense if the environment is unfavourable, in a positive sense if it is favourable. According to the ideas discussed in the last chapter, there is no question of the organism passively submitting to any such alteration. Instead, the critical point is the active adaptation of the organism to the environmental stress. The appearance of the inheritance of acquired characters occurs because of the selection of organisms which are better able actively to carry out an adaptation during their lifetime to the stresses impinging on them. Thus, the Soviet theory emphasizes a more or less passive response of the population as a whole, while the theory put forward here emphasizes in contrast the active ability of individuals to find some way of dealing with the situation.

However it is very important to realize that no such analogies between the biological processes of evolution and the conduct of human affairs should be taken very seriously. Certainly the rules of biological evolution apply to the biological evolution of man, but in relation to the general processes of human advance they cannot be regarded as more than suggestive. This is so because human advance does not take place only, or even mainly, by means of biological evolution. It is quite obvious that when the human species has developed a new capacity for dealing with its environment—for instance the ability to fly—it has not done so by means of the same kind of processes as were involved when other species of organisms, such as the ancestors of the birds, developed similar capacities. Biological evolution has in mankind been reduced to relative unimportance by the development of a new, and characteristically human, method of advance. We shall return to discuss the nature of this in a moment, but before doing so there is a further point to be made about evolution at the sub-human level.

There is another consideration about the general course of evolution which should have a very profound effect on man's general view of his place in nature, but which has in practice been unduly neglected. It concerns the type of things which undergo evolution. Biologists have always been under the strong temptation to regard their basic subject matter as composed of individual animals and plants, but one of the services of Darwinism, and particularly of the Mendelian neo-Darwinism of the last few decades, was to show that a more fundamental entity is the population. By this is meant a group of interbreeding individuals of the same species. A population, from the point of view of evolution theory, can be defined as a collection of individuals whose genes could be recombined to form the next generation. The transition from thinking in terms of individuals to thinking in terms of populations is another of those changes which has been hailed as a major advance in the conceptual scheme of biology. However, one can see now that like the 'biological relativity theory' proclaimed by Mayr, it did not go far enough. Populations of individuals of one species exist only as parts of a larger whole, which is the entire set of animals and plants which makes its living on a given part of the earth's surface. This whole system of living things is known as a biotic community. The study of the organization of communities, and the inter-relations of the various species which comprise them, is known as ecology. A fully comprehensive view of evolution must see it as a series of changes in communities, and in the types of organism which comprise them.

The study of communities essentially demands an organismic approach.* Perhaps the idea of a population, that is to say, a group of individuals of the same species, is indeed to some extent an atomistic notion—it is the kind of idea which is arrived at when one attempts to isolate separate parts within a community. But the ecology which studies communities as wholes cannot do other than concentrate its attention on connectedness and organization. The inter-relations within communities are in fact usually of great complexity. The fundamental theme around which they are organized is the winning and utilization of the energy which is required for organisms to maintain themselves. Ultimately, of course, this energy is derived from the sun. Any self-sufficient community must contain a basic stratum of organisms which are able to utilize directly the sun's energies to synthesize complex organic molecules; these are the green plants. On these plants many other organisms may live. The earliest theories of ecology supposed that it would be possible to exhibit the organization of a community in the form of comparatively simple 'food chains'. These would start with the primary producers which made direct use of solar energy; in the next level of the chain, the species b, c and d, would live by eating the primary producers; above them would be another level, m, n, o, which lived by eating b, c, d; and above them again perhaps a further level p, q, r, which lived by eating m, n, o; and so on. In practice, however, communities rarely have such a simple hierarchical structure as this. If, for example, grass is the primary producer, there will indeed live on this such grazing animals as cattle, sheep, etc., corresponding to b, c, d, and on these carnivores such as wolves and leopards, etc., corresponding to m, n, o; but in addition there will be worms, beetles, mice, etc., living on the grass, and on these another set of carnivores, such as weasels, which in their turn will provide some of the food for the wolves. There will be birds eating seeds as well as eating the insect grubs, others, such as owls, which live on some of the small mammals, and so on. The whole community, in fact, will form an extremely complicated network rather than a simple hierarchical pattern.

It is only rather recently that biologists have begun to come to realize the high degree of organization of the inter-locking relationships on which the workings of a living community depend. There are indeed very few types of community, even amongst the simplest ones which are found under the severest conditions of the desert or the arctic tundra, which have been fully described in detail. We are also still far from a proper theoretical understanding of community structure. It is interesting to realize that the mathematics required to express the structure of a community in which many populations of different species are related to one another is also applicable to many of the basic problems of development. In differentiation we are concerned with systems containing many different types of molecules, which compete with or influence one another in the utilization of the raw materials of growth, which in this case are simple chemical building blocks, such as the amino acids out of which proteins are built, or the nucleotides which form nucleic acids. From the formal mathematical point of view, the situation is essentially similar to that in a community in which many different species of organisms are competing for the energy which has been captured by the primary producers of the community. I have made some tentative efforts to utilize the mathematics of population growth, laid down by authors such as Lotka, Volterra and others, in connection with the theory of differentiation*; and quite recently a young mathematician in my laboratory, Brian Goodwin, has considerably developed this line of thought, and has produced a general system of mathematics, comparable to the statistical thermodynamics used in physics, which has applications to both the fields of community structure and differentiation.

It is a remarkable and regrettable fact that biological theory in this field is so comparatively backward. This is probably both a result and a cause of the widespread neglect of such ecological thinking in man's recent views about how his societies should be related to the material world he lives in. This part of biology should form a basic element in what used to be considered an essential aspect of philosophy, namely, economic and political philosophy. The application of scientific technology in human affairs has unfortunately produced results in this field which are not at all congruous with a truly scientific outlook. Since the end of the Industrial Revolution political philosophy, in the societies influenced by this revolution, has been unduly dominated by a climate of thought which may be appropriate to the simpler subject matter of physics and chemistry but which is exceedingly inadequate for the complexities of the human situation as it really is. The western industrial world has sought simply to utilize as much as possible of any sources of usable energy on which it could lay its hands. The sad story of the uncontrolled exploitation of the fertility of the prairies of the New World, which at one time reduced them to a dust bowl, is only a particular, rather extreme, example of the general way in which industrial man has been treating the world around him. There has been almost no realization that man is, and must always be, one element in an ecological community, and that if such a community is to continue operating successfully for long periods of time it must be organized with a carefully balanced system of checks and interactions between its various parts.

The development of a thorough understanding of ecological systems, and the application of such ideas to man's political and economic affairs, is nowadays rapidly becoming even more important. The major historical happening in the last half of the twentieth century is likely to be the taming for civilized life of the greater part of the earth's surface which is still, as we say, 'under developed'. In large parts of Asia, Africa and South

America the earth's surface has until now been utilized by man only for peasant subsistence agriculture. How can we turn tropical forests, Saharan savanna, or the African bush into suitable habitations for civilized and cultured men and women? It will certainly not be just by building a few automobile-dominated modern cities; nor can we rely in the long run on the extravagant exploitation of the stored-up energy in fossil fuels, such as coal and oil, on which western civilization has been largely based. We must find some self-sufficient system which produces as much energy as it uses. In time the production of energy from atomic processes will possibly solve this problem, but that date is likely to be some distance in the future. We need at least to approach the problem as an intelligent species, which, recognizing that it forms part of an ecological community, is considering how best that community can be organized. In the past we have, only too often, taken the attitude of a simple band of robbers, concerned only to get as much out of our surroundings as quickly as possible, with no thought of setting up a system capable of long-term operation. Perhaps the most important practical effect that the natural philosophy of biology could have the present time would be to show mankind a more truly scientific way of looking at his situation as an inhabitant of the world's surface.

However, the most important of all the lessons to be learnt by man from the consideration of evolution arises, not from the methods by which evolution has been brought about, but from the nature of the results which it has achieved. This lesson was in fact realized long before the theory of evolution was developed. It is the notion that the types of living things can be arranged in a hierarchy from lowest to highest. We have seen that this idea was well known to Aristotle, and was familiar to the Middle Ages in the form of the Great Chain of Being. In terms of evolution we express it by saying that there has been real evolutionary progress, which has led from the simplest and lowliest creatures by a series of stages to the most highly evolved type, which we consider to be man.

In quite recent years it has become fashionable in some circles to reject the idea that evolution has led to anything that is entitled to be called progress. This repudiation of the very ancient notion that some organisms are, in a meaningful sense, higher than others is probably largely a consequence of the reaction against the over-simple application of Darwinism to human affairs by the Social Darwinists. When it was argued that success in evolution went inevitably to those who were most ruthless in gaining advantage over their fellows during their lifetime, the idea of progress tended to become connected with the most blatant type of 'getting on'. Human progress seemed almost to be synonomous with imperialism and 'might is right'. When these political doctrines were repudiated, the concept of progress was abandoned along with them. But, as we have seen, there was never any real justification for associating worldly success with evolutionary fitness. A belief in the reality of evolutionary progress should never have been advanced as an argument for such political doctrines, and equally it should not be rejected along with them.

Evidence for or against the reality of progress is to be sought not in any implications which the idea may be thought to have for human affairs, but in the facts of biology. Many of the later products of evolution, such as the mammals in general, or man in particular, can do better many of the things that more primitive organisms do relatively inefficiently. For instance, many of the lowest organisms are stationary, like corals, or drift at the mercy of the currents in the sea, like jelly-fish, or move comparatively sluggishly, like many worms; while higher organisms can in general move much more actively and effectively. The contrast between higher and lower forms is even more striking nowadays, when we can seriously discuss the transition from non-living chemical molecules to systems which for the first time show the characteristics of life.

Evolution must have begun with the simplest possible systems which were able, firstly, to reproduce themselves, and, secondly, to undergo changes or mutations and then reproduce in the changed form. Such systems may perhaps have consisted of no more than a single type of molecule, perhaps of the nature of a nucleic acid, but it is perhaps more likely that the simplest conceivable system capable of undergoing evolution would require the presence of several different kinds of molecules. It seems rather probable that identical reduplication requires the presence of a nucleic acid, or conceivably some other complicated molecule which might fulfil a similar role of preserving some identity of structure through the process of multiplication; but it may well be that in addition to such information-preserving molecules some other types of chemical substance are required to carry out the actual work of forming the new material. However this may be, it is already clear that elementary assemblages capable of beginning a process of evolution are of extreme simplicity, when compared with such late products of evolution as the mammals. It is to the transition from these simple forms of life to the much more elaborate kinds which first appeared in relatively recent periods of the world's history that the name progress is given.

Those who object to the use of this name may agree that a series of changes have led on from the first organisms which began evolving to the more recently appearing forms, but argue that it is inappropriate to use the word progress to refer to it. Progress, they say, implies some form of betterment; and why should we consider ourselves in any way better than the worms? A short answer to this is similar to Dr Johnson's answer to those who queried the reality of the external world. He kicked his foot against a stone. We might say that we will take seriously the worm's claim to be our equals when the worms come and present it, but not before.

A more detailed reply can be found by examining the nature of the evolutionary processes and the kinds of results which they must be expected to produce. Let us consider some particular type of animal. As an example we may take the horses, whose evolutionary history is rather well-known. The recent members of the family are animals whose form of life involves nourishing themselves by eating grasses and relying on their fleetness of foot to escape marauding creatures which might prey on them, such as tigers or wolves. Their evolution will have been guided by the fact that those of them which left most offspring are likely to have been the animals which were most efficient in these respects. It is therefore only to be expected that we should find, as we do, that during their evolution the family of horses show signs of improvement in performing their characteristic way of life. Their jaws and teeth, for instance, have become larger and better suited to grinding hard grasses, and their legs have become longer and better fitted to enable them to run fast. Thus there has been something which is definitely an improvement within the terms of reference set by their particular way of life.

This is an example which shows that we may expect natural selection always to produce an 'improvement'; but it is important to realize that the changes it produces will be improvements from a particular point of view, and from other points of view may not be improvements at all. For instance, the changes in the horses's jaws are improvements from the point of view of eating grass, but they have rendered it almost impossible for the horse to eat certain other possible types of food, such as meat. Again, the changes in its limbs are improvements in producing a faster turn of speed, but this has been at the price of losing any possibility of developing manipulative skill. If, for instance, the climate were to change so that grass, instead of being a common plant became a rare one, the improvements which the horse family has undergone during its evolution would turn out to have led it into a dead end, from which it would probably not be able to escape; and the family would die out, so that its evolution came to an end. A fate of this kind has certainly overtaken many families of animals in the past.

When we turn from considering the evolution of one particular family of animals, with a characteristic way of life, to the inspection of the whole course of evolution from the simplest creatures onwards throughout the whole of the world, we are no longer provided with a simple criterion on which to judge improvement, since there is no longer any one mode of life against which to assess it. The living world as a whole is not faced by any one task, such as running fast and eating grass. The only enterprise on which it is engaged is the very general one of finding *some* way of making a living. But, just as the evolutionary processes will lead to an improvement in carrying out a particular task, so we may expect them to produce organisms which are improved in relation to this more general requirement. This is, in fact, what has happened; and it is improvement in respect of finding some way of exploiting the environment to make a living that justifies the use of the phrase 'evolutionary progress'. The groups of creatures that we speak of as 'highly evolved', are in general, less at the mercy of their environment than are the 'lower' organisms, and able to utilize, to carry on their life, more subtle relationships between environmental factors. Mammals, for instance, can keep their internal temperature constant, and thus attain a degree of independence of the surrounding temperature which is impossible to animals not able to do this. Again, it is only by the evolution of adequate organs of sight and scent and of a brain able to co-ordinate these, that the mode of living of a hunting wolf or tiger became a possible way of exploiting the surroundings to make a living.

We may conclude that the changes brought about by evolution will always be, in some sense, an improvement. In many cases they will be improvements only from a narrow point of view; and when circumstances change this may no longer be relevant, so that what had once been an improvement now leaves the animal defenceless to deal with the new situation. But, however many such setbacks may occur to particular families, the world of living things as a whole shows a continuous improvement in the ability to exploit, in some way or another, the surrounding circumstances so as to make a living off them.

This improvement is what we, quite justifiably, refer to as evolutionary progress. It is, as was said above, a fact of the greatest possible importance to mankind. It requires no great subtlety to recognize that the human species is still more efficient in finding ways of exploiting the environment so as to make a living, than any other species in the world. Man's abilities, in fact, carry forward the story of evolutionary progress into a new chapter.

The particular importance for man of the fact that evolutionary progress is a reality arises, however, from the peculiarity of man's nature and of the way in which his biological situation changes. Man has in effect produced a new mechanism which brings about alterations in his relations with the rest of the world as the generations pass. It is, if you like, a new method of evolution; but it operates by a method different from that on which biological evolution depends, and it might perhaps be questioned whether evolution is the right word to use in naming it. However, since there is no other word obviously available for this purpose, I shall continue to refer to the characteristic human process as 'human evolution', or, in the light of its mechanism, which we shall discuss in a moment, 'socio-genetic evolution'.

What the human species has developed is a new method of transmitting potentialities to later generations. This depends on the use of language. Items of fact, or methods of operation, can be *taught* by one generation to the next. Processes of teaching and learning carry out a function exactly analogous to that of biological heredity, in that they serve to specify the character of the new generation. This similarity in result can be indicated by referring to this method of passing on information as 'socio-genetic' transmission.

The development of what is in effect a new method of heredity must inevitably lead to the appearance of a corresponding new method of evolution. This method will, of course, not entirely supersede the biological type of evolution, which man undergoes like all living things, but will be supplementary to it. In addition to evolving biologically, man also evolves sociogenetically. Moreover, this socio-genetic evolution very rapidly results in changes which are, in human terms, of enormous importance. Within the period of recorded history we can detect only slight indications of biological evolution in human potentialities, but we are confronted with overwhelming evidence of most striking changes in human culture. These cultural alterations are not unidirectional, any more than were the changes produced by biological evolution through the long ages of animal life which led up to the appearance of man. Some cultures, like some animal species, have changed in directions which turned out to lead to a dead end, and some have altered in ways which from the general point of view must be considered regressions. But, just as in animal evolution as a whole we can see some direction which justifies us in speaking of certain groups as lower and of others as higher, so when we look at the whole of human cultural development we can see a general pattern of change, from small groups of nomads or scattered communities of food gatherers, to the complex and elaborate civilizations typified by such individuals as, let us say, Confucius, Plato, Newton and Leonardo.

It has often been argued that the existence, both within the sub-human animal world, and in the world of mankind, of general patterns of change which merit the title of evolutionary progress, provides us with an inspiration which should guide mankind's ethical strivings. One of the most prominent advocates of this type of ethical humanism at the present day is Julian Huxley. A similar argument has also been put forward from a more definitely religious point of view by Teilhard du Chardin. I personally agree very largely with their conclusions; but I should like to put forward an argument, which arises from a consideration of the nature of the human species as a biological entity, which I think establishes a still closer connection between human ethical beliefs and evolutionary processes. I have discussed this at some length in another book recently* and shall therefore only deal with the matter shortly here.

We need to consider carefully the requirements for any process which will make it possible for information to be transmitted from one generation to the next by teaching and learning. For such a process to be effective, it is not only necessary for a language to be developed in which the information can be expressed, but it is essential that the recipient should be brought into a frame of mind in which he is prepared to receive the information which is transmitted to him. In the process of biological heredity the new individual cannot avoid receiving the transmitted information, since half of it is already incorporated in the egg nucleus, and the other half is brought into the egg by the fertilizing sperm. In a process of cultural or socio-genetic transmission there must be some analogous function which ensures that the message is actually received. One might say, somewhat crudely, that socio-genetic transmission requires that first of all the recipient believes what he is told. If he does so, the transmission of information from one generation to the next is possible. Only after that, as a second stage, does the question arise of the person who has received the information comparing it with empirical observation, or with other items of information, and accepting or rejecting it for belief as we normally understand that word.

Recent studies on the psychological development of human infants, both by psycho-analysts and by authors such as Piaget, are beginning to throw some light on how this receptivity is formed in the developing mind. The human infant soon after birth seems not to be able to distinguish between itself and its surroundings. It exists, apparently, in a truly solipsistic universe in which it is the world and the world is part of it. The development of a mentality which is prepared to receive information transmitted from outside by language demands, of course, the breakdown of this solipsistic unity. The readiness to receive a transmitted message implies that there appears within the infant's mind some psychological system which carries the authority which is necessary for the information not only to be taken in, but to be allowed to have meaning. Now it appears, as a fact of empirical observation, that the development of the authority-bearing system required for the transmission mechanism goes on at the same time as, and in intimate conjunction with, the formation of the notion that some things are good, and others bad, in an ethical sense. It is, of course, a major characteristic of ethical ideas that they involve some sort of authority. I do not mean that our ideas of good and bad are necessarily imposed on us by some outside authority, such as that of our parents, or of the church. What I mean to say is that it is an essential part of the idea of the ethically valuable that it is authoritative in the sense of tending to impose an obligation.

The argument I wish to advance is that the authority which is required to make possible the socio-genetic method of transmitting information, and the authority which is involved in the development of the ideas of ethical good and bad, are two aspects of one and the same type of mental functioning. It might perhaps be possible to conceive, in theory, of a method of cultural transmission which involves some form of authority not of the kind which leads to the development of ethical beliefs; but in fact the mechanism which has actually been produced during the evolution of mankind is one in which these two aspects of authority are indissolubly connected.

If this is the case, the fact that man is the sort of creature who goes in for having ideas of right and wrong is an essential part of the same mechanism which makes it possible for him to transmit information by teaching and learning. The basic nature of our ethical character is, then, that it is a part of our special human or socio-genetic type of cultural hereditary mechanism. It would follow from this that, just as we can judge genetic changes by whether they are suitable to carry forward evolutionary progress on the biological level, so we can judge various different types of ethical belief according to whether they seem likely to carry forward human evolution on the cultural level. Any understanding that we can attain of the general pattern of human cultural evolution provides us with a criterion of what one might call 'wisdom', which can be applied to judge whether one type of ethical belief is to be preferred to another.

In this argument I have applied to human affairs a kind of reasoning which is often used in biology. In dealing with living organisms we are frequently faced with the problem of defining a criterion by which particular functions can be judged. It is a commonplace in biology to discover such criteria by an adequate examination of the material which can be empirically observed. In this way, for instance, we can form a concept such as that of health, or normal growth. We have no hesitation in using such concepts to judge between, for instance, different diets. I am suggesting that the concept of evolutionary or cultural progress has a similar degree of validity to the concept of health, and can be used in a like manner to judge between different examples of one particular human function, namely the function of holding ethical beliefs.

Whether this is so or not—and for a further discussion of this complicated question I should like to refer you to my recent book on the subject *The Ethical Animal*—no one will deny that the fact that man is a creature who goes in for entertaining ethical beliefs, is one of his most important characteristics. It confronts us with a problem which has always been recognized as one of the most difficult which natural philosophy has to tackle. That is the old conundrum of free-will. For a clear statement of the point we can go back again to *Paradise Lost.* In a later part of Raphael's speech to Adam, the archangel says:

> God made thee perfect, not immutable;
> And good he made thee; but to persevere
> He left it in thy power—ordained thy will
> By nature free, not over-ruled by fate
> Inextricable, or strict necessity.
> Our voluntary service he requires,
> Not our necessitated. Such with him
> Finds no acceptance, nor can find; for how
> Can hearts not free be tried whether they serve
> Willing or no, who will but what they must
> By destiny, and can no other choose.

The difficulty, of course, is that our will is obviously connected with a material structure, namely, our brain and the events going on in material structures proceed according to certain laws of causation. If causation is strictly deterministic how can there then be any freedom of the will?

One of the most influential trends in recent philosophy has been the attempt to show that many of the long-standing puzzles, by which philosophers have felt themselves defeated, have in fact only arisen from a misuse of language, and are actually senseless, that is to say, do not require, and in fact could not possibly receive, an answer. Some of the recent followers of the school of logical positivism, which has been particularly active in this way, have argued that the difficulty about free will is of this kind. Perhaps the most important such statement is that by Professor Ryle in his book *The Concept of Mind.* He advances two main points, but they seem to me both inconsistent with one another and unconvincing in themselves when taken separately.

The first argument purports to show that the laws of physics 'may, in one sense of the metaphorical verb, govern everything that happens, but they do not ordain everything that happens'. They thus leave room for the occurrence of processes which are not governed by strict determinism. Ryle attempts to establish this by means of an illustration. Consider a scientifically trained spectator who is observing a game of chess. He would after some time be able to deduce laws which govern the moves which the various men can make. These rules would always remain valid; but it would also be true that they do not ordain what happens in the game, since every game is different from every other. But this surely misses the entire point, which only arises when we suppose that there are laws which control, not merely the moves which can be made with the chessmen on the board, but the material happenings in the brains of the players. Ryle has put forward, as a model of the natural universe, a system which includes players who have free-will. Obviously enough, within such a system there is no problem as to whether free-will can exist or not, since the author has already put it there.

But having thus made room for the existence of processes of an undetermined kind, Ryle proceeds to argue that it is unnecessary to refer to them. People he says 'often pose such questions as "how does my mind get my hand to do what my mind tells it to do?"' (He is thinking of someone pulling the trigger of a pistol.) He goes on 'Questions of these patterns are properly asked of certain chain-processes. The question "What makes the bullet fly out of the barrel?" is properly answered by "The expansion of gases in the cartridge" '; the question 'What makes the cartridge explode?' is answered by reference to the percussion of the detonator; and the question 'How does my squeezing the trigger make the pin strike the detonator?' is answered by describing the mechanism of springs, levers and catches between the trigger and the pin. So when it is asked 'How does my mind get my finger to squeeze the trigger?' the form of the question presupposes that a further chain-process is involved, embodying still earlier tensions, releases and discharges, though this time 'mental' ones. But whatever is the act or operation adduced as the first step of this postulated chain-process, the performance of it has to be described in just the same way as in ordinary life we describe the squeezing of the trigger by the marksman. Namely we say simply 'He did it' and not 'He did or underwent something else which caused it.'

But surely in fact we do not remain content merely to say 'he did it', we proceed to ask whether he did it because somebody jogged his elbow or pricked his arm with a pin which caused an involuntary contraction of his muscles, or whether in fact he pulled the trigger voluntarily by an act of will. It is not sufficient simply to shut one's eyes to such questions and refuse to ask them. Deciding to do something and then doing it is a common experience. You know perfectly well that you can at this moment stop listening to me and start attending to the frogs which are singing their usual songs in the tropical night outside this lecture room; or you can decide to go on listening to me. You have, I think, a very strong experience that you are free to do one or other of these as you wish. This experience of what seems to be freedom cannot be simply left out of account as something that does not occur. It demands some sort of an explanation.

Another direction in which people have recently sought for a way of understanding this problem is in relation to the changes which have been taking place in our conceptions of physical causation. Physicists no longer consider that the interactions between the ultimate particles of matter take place according to the strict laws of determinist causation. They envisage them instead as occurring according to statistical laws, or laws which involve a certain degree of 'indeterminacy'. Does this abandonment of the older ideas of causation create, as it were, an area of freedom in which free will could co-operate?

When the principle of indeterminacy was first established there was some tendency to believe that it might in fact provide a way of escaping from the dilemma about free will, but most people who have considered the matter carefully have by now come to the conclusion that it does not do so. The indeterminancy applies to interactions between elementary particles. These are events of almost indescribably small dimensions. Any single nerve cell of the human brain contains vast numbers of atoms or electrons; and as soon as large numbers are involved the degree of indeterminacy becomes less. If you toss a penny once it is quite undetermined whether it will come down heads or tails, but if you toss a thousand pennies there is very little likelihood that the result will differ much from five hundred heads and five hundred tails; if you toss a million pennies the total result is even more nearly determined. In any material system large enough to play an important part in the functioning of the brain there must be so many elementary particles that the results of their interaction must be almost completely determined. Scarcely any 'play' will be left in the causal mechanism, in which free-will could operate.

It is perhaps not impossible that ways will be found for avoiding this conclusion. For instance, J. B. S. Haldane* once made the ingenious suggestion that mind is a resonance phenomenon. He pointed out that nerve cells undergo periodic electric disturbances, and supposed that mental phenomena, such as awareness or will, are associated with these disturbances being in time or in phase with one another. Now, resonance phenomena involve very small amounts of energy, and thus the degree of indeterminacy associated with them might be quite large. It is conceivable that, by some such hypothesis as this, we might show that some mental phenomena are, after all, of a kind for which the indeterminacy principles of modern physics are important.

However, even if our theories of the brain ever did develop to this point, should we in fact be any further forward in accounting for the experience of free-will? It is true that we should have provided for there being some play in the mechanism by which past events determine the future ones, and should to that extent be no longer confined in the strait jacket of strict determinism. But, in any strictly physical system, this play would be taken up by the operations of chance, according to statistical laws. The experience of free-will is not that of some rolling of dice within our minds; it is, exactly, that we can *choose* one result rather than another, and not that the choice has to be left to chance. Thus, if we look to modern physics to provide a solution to the problem of free-will, we should have to rely not only on the principle of indeterminacy but on a Maxwell's Demon, that is to say a spirit who can control the operations of chance for us.

The real point, surely, of the experiences to which we give the name of the exercise of free-will, is that they involve not only determining what will happen in the future, but the exertion of some kind of effort to achieve this result. It is when we are tempted to do something else that we experience most keenly the feeling that by an 'effort of will' we can decide what action we shall really take.

One aspect of our mental life which seems to be very relevant in this connection, but which is not often mentioned in discussions of the problems of free-will, has been brought into considerable prominence in recent psychological work. That is the fact that apparently all types of activity which go on in our brain can proceed unconsciously as well as consciously. It has, of course, long been realized that the performance of such tasks as preserving our balance while walking involves the passage of nervous impulses from and to our muscles, and their combination within the brain, by processes of which we are normally quite unaware, but of which we can become conscious if an injury to a limb or another unusual situation makes this necessary. The investigation of the unconscious by psychoanalysts and others seems to demonstrate that other types of mental events, which might have been thought much more indissolubly linked with consciousness, in fact need not be so. Impressions can be received by our sense organs, and affect the brain in such a way that they can be later recalled in memory, even when we were quite unconscious of perceiving them. We seem here to be confronted with all the brain-events involved in perception, but without the self-awareness. Again, it has become commonplace to speak of unconscious willing or unconscious desires, which again suggests that the nervous processes associated with these types of mental activity may proceed without awareness being attached to them.

If this is so, the experiences to which we give the name of free-will cannot depend wholly on the particular type of nervous activity which, when it is expressed in action, appears as a purpose, but must essentially involve a phenomenon of self-awareness in addition to this.

The existence of unconscious willing, if it is accepted, raises a number of interesting questions for philosophy. Is an unconscious will necessarily teleological? Natural philosophy nowadays rejects teleological ideas because they appear to demand the existence of some self-aware being who can formulate purposes and ends. If there can be unselfconscious acts of willing this difficulty may no longer arise. The whole matter seems to require restatement in terms of the newer concepts. One attempt to do this has been made by Sinnott.* He advanced the argument 'that biological organization (concerned with organic development and physiological activity) and psychical activity (concerned with behaviour and thus leading to mind) are fundamentally the same thing. This may be looked at from the outside, objectively, in the laboratory, as a biological fact; or from the inside, subjectively, as the direct experience of desire or purpose. . . . It is this organization, whatever it may turn out to be in terms of matter and energy, space and time, which, as experienced by the organism, I believe to be the simplest manifestation of what in man has become conscious purpose. Just as the form of the body is imminent in the egg from which it grows, so a purpose, yet to be realized, may be said to be imminent in the cells of the brain.'

This is, in effect, to bring the formation of purpose into relation with what we in an earlier chapter referred to as creodes, that is to say, self-regulating processes of change which are so organized that they tend to arrive at a determined goal. As we have said, systems of this kind, organized on the basis of feed-back and cybernetic relations, have what we called a quasi-finalistic character. A gun which incorporates a target-tracking sight might indeed be regarded as a system embodying an unconscious purpose. When Sinnott wrote he did not refer to the newer ideas of cybernetics or feed-back systems, but they were clearly implied, and the attention which has been paid to them recently only makes more plausible and productive the point of view which he was putting forward. It is today not too difficult to conceive, in general terms, of ways in which the nervous activities in the brain might be organized into systems which would function in a manner which we would recognize as the operation of an unconscious will. We are getting to the position where we require a new vocabulary to discuss such systems. It is really hardly satisfactory to speak of the hitting of a target as the 'unconscious purpose' of an automatic gunsight, or even as its goal. It would be better to have some new technical and neutral term, less closely involved with earlier preconceptions.

We must return, however, to the point that the experience of free-will essentially involves self-consciousness. Now this is not quite the same thing as saying that it is only to the conscious exercise of free-will that we can attach moral praise or blame. One can conceive of a person who performs good actions spontaneously, naturally and unself-consciously, in fact the formation of such personalities has been the ideal of some schools of Oriental philosophical thought such as Zen Buddhism. They are people whose mental creodes (i.e. unconscious purposes) are directed towards ends which we recognize as good and such people are considered as worthy of moral approval. The question of whether there is awareness of purpose therefore does not necessarily arise in the moral connection. The possibility of attributing moral praise or blame depends only on the existence of purposes, whether conscious or not, and on there being a 'freedom' which makes it meaningful to say that there is a choice between purposes or that a purpose can have an effect on the action carried out.

The importance of self-awareness in this connection is that the only evidence we have that such freedom exists is our conscious experience of performing acts of will. This evidence is, in my opinion, impossible to interpret at the present time, simply because we have no way of interpreting self-awareness in terms of anything else. We are aware, in the form of visual sensations, of some of those activities of our brain cells which are set in action by light waves impinging on the retina; we remain unaware of other similar nervous activities which are set going by other light waves impinging at the same time. The evidence that these 'unconsciously perceived' nervous events could be recoverable from memory later on leaves us wholly in the dark as to the nature of the difference between the events we are aware of as sensations and those of which we remain unaware. Moreover, self-awareness is a phenomenon of a kind for which it seems impossible to see how any explanation in terms of observable phenomena could ever be provided. In however much detail we could describe the electrical phenomena going on in our brain cells, this description could never have as its consequence that an act of self-awareness occurred. Awareness can never be constructed theoretically out of our present fundamental scientific concepts, since these contain no element which has any similarity in kind with self consciousness.

Some thinkers have concluded from this that, since self-awareness undoubtedly exists, the mode of it which we experience must have evolved from simpler forms which are experienced by non-human things. It is not at all difficult to accept this about other non-human animals. In fact, observing a dog or a cat going to sleep and waking up, we normally, I think attribute to it some form of self-consciousness, of a roughly similar kind to that which we know in ourselves. But if we take the theory of evolution seriously, where are we to stop? Was there a point on the evolutionary scale at which self-awareness first originated? But, as we have seen, it is inconceivable that it originated from anything which did not share something in common with it but possessed only those qualities which can be objectively observed from outside. Are we not forced to conclude that even in the simplest inanimate things there is something which belongs to the same realm of being as self-awareness? It need not, of course, resemble our self-awareness any more closely than say, the passage of an electric current down a wire resembles the operation of a complex calculating machine, or the operations of the nerve cells in our brain. But, just as both a simple electric current and the operations of a computer can be described in terms of electrical circuits, so, according to this line of thought, something must go on in the simplest inanimate things which can be described in the same language as would be used to describe our self-awareness.

Such ideas have been expressed recently by Teilhard du Chardin,* who maintains that, as well as what he calls the 'without' of things, which is what we can observe, there is always a 'within' which we cannot observe except in ourselves, but which always has something of the quality of self-awareness. I cannot myself follow all the conclusions that du Chardin draws from this, but the basic idea finds support from many quarters. For instance, du Chardin quotes J. B. S. Haldane: 'We do not find obvious evidence of life or mind in so-called inert matter, and we naturally study them most easily when they are most completely manifested; but if the scientific point of view is correct, we shall ultimately find them, at least in rudimentary form, all through the universe.'

For Whitehead the phenomenon of self-awareness is connected with organization. It is obviously highly developed in man, who is the most highly organized of living things. Whitehead* suggested that something of the same general nature as self-awareness might be highly developed also in some of the simpler inanimate things which exhibit a high degree of organization, such as atoms. In between the world of atomic physics and that of biology is a realm of supra-molecular aggregations of matter, in which 'the organic unity fades into the background'. 'If we wish to throw light on the facts relating to organisms', Whitehead writes, 'we must study either the individual molecules and electrons or the individual living beings; in between we find comparative confusion.' These are, of course, highly speculative notions. The idea that something of the same general kind as self-awareness may exist in inanimate systems is one which we may feel is forced upon us by the demands of logic and the application of the evolutionary theory, but it is quite clear that even if this self-awareness exists in atoms and electrons we know nothing of its nature. It does not provide us with any understood antecedents from which we can gain any insight into the ultimate nature of our own self-consciousness, or its relations to our apparent experience of free-will.

We confront, in the phenomenon of self-awareness, a basic mystery which lies at the heart of our whole life. It is not only the experience of free-will which is inextricably involved with self-awareness; our whole understanding of the external world is deduced from what we consciously perceive. We can explain to ourselves some of the mechanisms by which, for instance, light waves emitted from an object are focused on our retina, and there cause electrical disturbances which travel into and around our brains. But the most essential step in the whole process is that these events cause us to be aware of something; and so long as we have not the faintest idea what this awareness means, and cannot envisage any way in which the phenomenon of awareness could be expressed in terms of anything else, the act of perception, and the whole observable world which depends on it, contains an inescapable element of mysteriousness. The sciences of physics and chemistry can at least aspire to give a coherent account of the world with which they deal because their interests are confined to a comparatively narrow field. Biology cannot avoid having to discuss the nature of man, and one of the most obvious facts we know about man is that he is aware of himself. Since the nature of self-awareness completely resists our understanding, the natural philosophy of biology cannot but bring us face to face with the conclusion that, however much we may understand certain aspects of the world, the very fact of existence as we know it in our experience is essentially a mystery.

I suppose you will probably expect me to end there. For a scientist to point out that the world is after all, in spite of all his telescopes and microscopes, experiments with atomic particles and physiological apparatus, a highly mysterious place—surely that is enough to persuade the poets and artists that the scientist is after all quite a profound fellow, or at least not so shallow as they had thought him. But, actually there is something more to be said.

It is true that perception, even by means of the sense organs whose workings we understand in some detail, such as the eye or the ear, is a fundamentally mysterious process, since it depends on a phenomenon of self-awareness which we do not understand intellectually at all. It cannot be denied therefore that there might be other ways of arriving at something which would be justifiably called a knowledge of the existent world; for instance, by intuition or some other form of mystic insight. But the biologist would be doing less than justice to his science if he failed to point out that man's methods for apprehending his surroundings must have been evolved like his other functions, and that this evolution must have depended on selection for some kind of efficiency of operation. Darwin himself was impressed by this. In his mind it raised a feeling of doubt and uncertainty which I do not myself experience so strongly.

Darwin, as the pioneer of evolutionary thinking, was only just escaping from the grip of the old presupposition that knowledge should be totally objective, without any flavour derived from man's own nature; he was also obsessed with the task of raising every possible objection to his theory of evolution, and either refuting it, or at least showing that, in spite of being aware of a difficulty, he felt that mature judgment would still come down on his side. From the argument that man's mentality is shaped by evolution, he was led to ask whether there could be any final validity in the picture which a mind formed in this way could draw of the processes by which it had itself come into being.* Can a product of evolution, he asked, arrive at a true conception of evolution? I think that nowadays we would be tempted to answer that of course such a mind cannot produce a theory which is completely objective and uninfluenced by the mind's own nature; but we should go on to argue that there is no sense or meaning in aspiring to total objectivity. We are a part of nature, and our mind is the only instrument we have, or can conceive of, for learning about nature or about ourselves. The idea that during evolution the mentality of animals becomes shaped by its use should be no more a reason for mistrust than is the use-determined curvature of the handle of a scythe or the shape of a knife blade. If there are other types of cognition, not dependent on sense awareness and conceptual thought, the fact that they are so little developed in the majority of mankind indicates either that they are less efficient in the carrying out of man's essential biological activities, or that they are beyond his present capacities. If man has not yet evolved the capacity to exercise efficiently this hypothetical other method of cognition, then it is, perhaps, worth striving for; but there is no reason to suppose that in order to attain it we should have to abandon the conceptual and perceptive mental abilities which we now possess. If, on the other hand, these other possible methods of cognition have been tried and found less effective in the past—for instance, in the alleged telepathic and intuitive powers sometimes attributed to various primitive peoples such as the Australians—again we should have to conclude that our conceptual form of intellectation, of which science is the highest form, must remain the major type of cognition of which we should rely.

We might, of course, come to the conclusion that although other forms of cognition were less efficient they still potentially have an important part to play, and that mankind had gone too far in rejecting them. It is perfectly possible, indeed perhaps advisable, to regard modern western civilization as in danger of too great specialization, one aspect of which would be its too great reliance on a conceptualized form of self-awareness. But even if we adopt this point of view, we should be led to suggest only a comparatively slight redistribution of emphasis as between the types of cognition of which man is capable. The basic fact would remain that our existing faculties of perception and conceptual thought have been brought into being by a process of evolution which involves passing the rigorous test of natural selection. Our realization of the essential mysteriousness of the universe does not stand alone; it is complemented by our confidence that the faculties we have for dealing with our surroundings must have at least a considerable degree of effectiveness.

NOTES AND REFERENCES

THIS little book is, of course, in no sense a text-book or work of reference. Full citations of the sources of the facts and ideas discussed would therefore be superfluous. I shall mention a few general texts which can be used for further reading or as guides to the literature. Otherwise references will be provided only for items which seem to call for it for some special reason, either because they are very recent, or because they are suggestions of my own which may not be fully accepted into the general consensus of biological opinion and which some readers may wish to examine in more detail.

p. 28. Standard recent text-books of genetics are: Sinnott, Dunn and Dobzhansky, *Principles of Genetics*, New York, McGraw, 1958. Srb and Owen, *General Genetics*, Freeman, San Francisco, 1952.

A very clear account of the elements of the subject is given by Philip Goldstein, *Genetics Made Easy*, London, Rider, 1957.

p. 36. Waddington, *Introduction to Modern Genetics*, London, Allen & Unwin, 1939.

p. 38. For recent work on the fine structure of genes and its relation to nucleic acids and proteins, see Pontecorvo, *Trends in Genetic Analysis*, Oxford University Press, 1959; *The Chemical Basis of Heredity* (edited by McElroy and Glass), Oxford University Press, 1957; *Biological Organization* (edited Waddington), London, Pergamon Press, 1959. A more elementary account will be found in the *Scientific American* book, *The Physics and Chemistry of Life*.

p. 45. A good recent general discussion is Brachet, *Biochemical Cytology;* for the change from embryonic to adult conditions, see Waddington and Sirlin, *Exp. Cell. Res.*, 1959, *17*, 582.

p. 50. Fincham, J. R. S. and Pateman, J. A. *Nature*, London, 1957, *179*, 741.

p. 50. Lewis, E. B., Cold Spring Harb., *Symp. Quant. Biol.*, 1951, *16*, 151.

p. 53. General textbooks on the study of development are: Balinsky, *An Introduction to Embryology*, London, Saunders, 1960; Waddington, *Principles of Embryology*, London, Allen & Unwin, 1956; Willier, Weiss and Hamburger, *The Analy-*

sis of Development, London, Saunders, 1955; Kühn's *Entwicklungsphysiologie* is very good, but only available in German.

p. 57. The ideas discussed in the next few pages were first introduced in Waddington and Schmidt, Roux Archiv. 1933, *128*, 522. They are discussed at some length in my books *Organizers and Genes*, Cambridge University Press, 1940, and *Epigenetics of Birds*, Cambridge University Press, 1952, and in Needham's *Order and Life*, Cambridge University Press, 1936, and *Biochemistry and Morphogenesis*, Cambridge University Press, 1942.

p. 62. Kroeger, H., *Chromosoma* 1960, *11*, 129.

p. 63. The fact that there are only relatively few alternative paths of differentiation open to cells began to be emphasized by embryologists in the 1930s (Cf. Weiss, *Principles of Development*, Holt, New York, 1939). The attempt to interpret these paths as the results of the interactions of gene-controlled activities was begun in my *Organizers and Genes*, and is discussed more fully in my *Strategy of the Genes*, London, Allen & Unwin, 1957.

p. 67. Surprisingly few studies have been made in which cells have been examined with the electron microscope during their alteration from the embryonic to the fully differentiated condition. The ideas in the text are derived from the facts recorded in Waddington and Perry, *Proc. Roy. Soc.* Lond., B., 1960.

p. 68. See Mookerjee, Deuchar and Waddington, *Journ. Embryol. Exp. Morph.*, 1953, *1*, 399; and Waddington and Perry (in press).

p. 70. Saunders, Cairns and Gasseling, *Journ. Morphol.*, 1957, *101*, 57.

p. 72. Some outstanding accounts of modern evolution theory are: Huxley, *Evolution: the Modern Synthesis*, London, Allen & Unwin, 1958; Dobzhansky, *Genetics and the Origin of Species*, Oxford University Press, 1951; Simpson, *The Meaning of Evolution*, Oxford University Press, 1950. Many of my own ideas expressed in this chapter are more fully discussed in *The Strategy of the Genes*.

p. 74. Willey, *Darwin and Butler: Two Versions of Evolution*, London, Chatto & Windus, 1960.

p. 75. See Preface to Darwin's *Voyage of the Beagle*, edited Nora Barlow, Cambridge University Press, 1933.

p. 78. See Nicolson, *Newton Demands the Muse*, Oxford University Press, 1946, and Lovejoy, *The Great Chain of Being*, Oxford University Press, 1936.

p. 79. Cannon, *Lamarck and Modern Genetics*, Manchester University Press, 1959.

p. 83. Yule, U., *New Phytologist*, 1902, *1*, 193.

p. 84. Mayr, *Where are We?* in *Cold Spring Harb. Symp. Quant. Biol.*, 1959, *24*, 1.

p. 86. Quoted by Basil Willey, *loc. cit.*

p. 86. Irvine, *Apes, Angels and Victorians*, London, Weidenfeld & Nicolson, 1955.

p. 87. Fisher, R. A., *Creative Aspects of Natural Law*, Cambridge University Press, 1950.

p. 90. Kettlewell, *Nature*, 1955, *175*, 943.

p. 93. The ideas developed in the next few pages were first adumbrated in *Nature*, London, 1942, *150*, 563. Several recent experiments are discussed in my *Strategy of the Genes*. The experiment with salted food appeared in *Nature*, London, 1959, *183*, 1654. See also *Evolutionary Adaptation*, in *Evolution After Darwin*, vol. 1, edited Sol Tax, Cambridge University Press, 1960, and 'Evolutionary Systems—Animal and Human', *Nature*, London, 1959, *183*, 1634.

p. 102. For a good discussion of ecology in relation to evolution, see Marston Bates, *The Nature of Natural History*, Scribner, New York, 1950; and also his article, 'Ecology and Evolution' in *Evolution After Darwin*, vol. 1, edited Sol Tax, 1960.

p. 103. *Strategy of the Genes.*

p. 111. *The Ethical Animal*, London, Allen & Unwin, 1960.

p. 117. Haldane, *Philos. Sci.*, 1934, *1*, 78.

p. 119. Sinnott, *Cell and Psyche*, Oxford University Press, 1950.

p. 121. du Chardin, *The Phenomenon of Man*, London, Collins, 1959.

p. 122. Whitehead, *Science and the Modern World*, Cambridge University Press, 1926, p. 156.

p. 124. For another expression of this point of view, see Schroedinger, *Mind and Matter*, Cambridge University Press, 1958.

Index

For some often recurring words, e.g. chromosome, a reference is given only to one or a few places where its meaning is defined.